D0626819

ELEMENTS

of

STEREOCHEMISTRY

ERNEST L. ELIEL

Professor of Chemistry
University of Notre Dame

with a section on coordination compounds by

F. BASOLO,

Professor of Chemistry
Northwestern University

JOHN WILEY & SONS
New York / London / Sydney / Toronto

PREFACE

This brief and elementary summary of the subject of stereochemistry was originally composed for the benefit of the non-chemist. Alas, and perhaps justifiedly, it was not considered elementary enough for this purpose. However, several of my colleagues have urged me to arrange for its publication in booklet form for the benefit of the beginner in organic chemistry.

Stereochemistry has become one of the cornerstones of organic chemistry. Several (though unfortunately by no means all) modern textbooks of organic chemistry have taken note of this fact by introducing stereochemistry very early. This has had the beneficial result of permitting virtually all organic chemistry to be explained and interpreted on a basis which includes stereochemistry. On the other hand, it has sometimes had the drawback of leading to a fragmentation of the subject; as a result, the student is at times not satisfied in his quest for a clear and comprehensive understanding of stereochemical concepts and terminology.

The present booklet hopefully provides for that understanding. It is meant to be used in conjunction with a textbook of basic organic chemistry and it is not meant to supplant existing more detailed textbooks of stereochemistry. Such textbooks are designed for the more advanced student and for the occasional exceptionally brilliant beginner. This book is written for any student capable of following a course in organic chemistry; hopefully it will inspire a few to go on to a more thorough study of a topic which is of fundamental importance in subjects ranging from structural chemistry to molecular biology. A brief section on inorganic chemistry is included to make it clear, even to the beginner, that the importance of stereochemical concepts is not confined to organic compounds. The presence of this section should also make it possible for Departments which no longer employ the traditional divisions of chemistry (inorganic, organic, physical, etc.) to use the present book. The section on inorganic stereochemistry was written by Professor F. Basolo of Northwestern University to whom I am grateful for this important contribution.

I also wish to express my special thanks to Dr. Hardy Christen and the Juris-Verlag of Zürich, Switzerland for having produced a manuscript suitable for offset printing, and to Professor G.F. Hennion (University of Notre Dame, Indiana), and Dr. R. Scheffold (University of Fribourg, Switzerland) for a critical reading of the manuscript and many helpful suggestions.

June 1968   Ernest L. Eliel

CONTENTS

I. INTRODUCTION

The term stereochemistry (from Gr. "stereos" = solid + chemistry) refers to the three-dimensional architecture of molecules (i.e. the arrangement of the constituent atoms in space) and to that part of chemical behavior which is dependent on this arrangement ("dynamic stereochemistry"). In determining the structure of any substance--for example lactic acid which is found in sour milk--the first step is to determine the molecular formula, i.e. the kind and number of the atoms which make up the molecule. In the case of lactic acid this formula is $C_3H_6O_3$. The next step is to determine the chemical "constitution", i.e. to find out which atoms are linked to which others; this yields the formula

$$
\begin{array}{ccc}
\text{H} & \text{H} & \\
| & | & \nearrow\text{O} \\
\text{H-C-C-C} & & \\
| & | & \searrow\text{O-H} \\
\text{H} & \text{O-H} & \\
\end{array}
$$

However, the information conveyed by this formula is still incomplete; it turns out that there are two kinds of lactic acid, that found in souring milk ("fermentation lactic acid") and that produced in living muscle when it performs work ("sarcolactic acid"). In their physical and chemical behavior, the two lactic acids are extremely similar (for example, they have the same melting point, vapor pressure, density, refractive index, acid strength, infrared spectrum, nuclear magnetic resonance spectrum and reactivity towards ordinary chemical reagents). However, they differ in their action on polarized light (see below), one rotating the plane of polarized light in a clockwise direction (sarcolactic acid or "(+)-lactic acid") and the other rotating such light in a counterclockwise direction (fermentation lactic acid, or "(-)-lactic acid"--the sense of rotation here is defined as that seen by an observer toward whom the light is traveling). More significantly the lactic acids differ in their biological (or biochemical) behavior; thus sarcolactic acid, (the "dextrorotatory" isomer, i.e. the one rotating to the right) but not fermentation lactic acid (the "levorotatory" or left-rotating isomer) is dehydrogenated to pyruvic acid,

$$
\begin{array}{ccc}
\text{H} & \text{O} & \\
| & \| & \nearrow\text{O} \\
\text{H-C-C-C} & & \\
| & & \searrow\text{O-H} \\
\text{H} & & \\
\end{array}
$$

in the presence of the enzyme lactic acid dehydrogenase isolated from beef heart.
There are innumerable examples of such biochemical differences; we will mention here only the difference between (+)- and (-)-asparagine (only the + or
dextrorotatory isomer is sweet), between natural (+)- and unnatural (-)-glutamic
acid (only the dextrorotatory compound functions as the familiar favor-enhancing
agent) and between the "stereoisomers" (see below) of chloroamphenicol (chloro-
mycetin) of which only one acts as an antibiotic. (Here, for reasons to be
explained later, there are four stereoisomers, three of them pharmacologically
inefficacious.)

Clearly the cause or basis, at the molecular level, of the difference between
dextrorotatory and levorotatory lactic acid (or similar pairs of compounds of
opposite rotation) is of great importance and has thus generated considerable
interest. Le Bel and van't Hoff, following earlier speculations by Pasteur (see
below) proposed the now accepted explanation, namely that the difference between
the lactic acids is due to a different arrangement in space ("configuration") of
the atoms. It is now generally accepted and amply supported by physical as well
as chemical evidence that the valencies of saturated carbon are tetrahedrally
disposed; thus the molecule of methane, CH_4, is properly described by a tetrahedron with the hydrogen atoms at the corners and the carbon atom at the center
(Fig. 1)

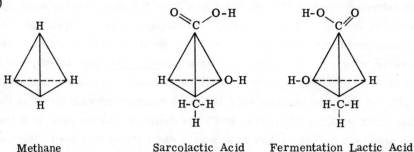

Methane   Sarcolactic Acid   Fermentation Lactic Acid

Fig. 1

When the four atoms or groups attached to the central carbon atom are all different
as in lactic acid (CH_3, H, OH and $C{\lessgtr}^O_{OH}$) two arrangements are possible, as shown
in Fig. 1. These arrangements (best appreciated in a molecular model) are not
superposable (i.e. non-identical) but are mirror images of each other; they
correspond to the dextrorotatory and levorotatory forms of lactic acid and are

said to be opposite in their "configuration", though identical in constitution.

Since the two lactic acids have the same number and kind of atoms (formulas $C_3H_6O_3$) they are isomers and since the difference between them is in their three-dimensional molecular architecture (arrangement of atoms in space) they are called "stereoisomers" and are said to differ in stereochemistry.

The fact that molecules may differ in configuration (stereochemistry) and that this difference produces differences in chemical and biochemical behavior as well as in optical rotation raises a number of fundamental questions, such as the following: Why do some substances rotate the plane of polarized light and not others? How are configurational isomers best represented and how are they named? Why do they differ in some chemical properties and not in others? How are differences in certain physical properties (e.g. optical rotation) related to configuration? What configuration or spatial arrangement corresponds to which isomers? (e.g. are the stereoformulas assigned to the two lactic acids in Fig. 1 correct or should they be inverted), etc. The paragraphs following the brief historical survey will address themselves to these questions.

II.   HISTORY AND BACKGROUND

Optical rotation was discovered by Etienne L. Malus in 1808. When ordinary light (whose electromagnetic waves may occupy any one of the infinite number of planes intersecting in the direction of propagation of the light ray) is passed through a crystal of Iceland spar (crystalline calcium carbonate) two rays are produced by a phenomenon called "double refraction"; these two rays are "plane polarized" at right angles to each other, meaning that in each case the magnetic wave associated with the light ray vibrates in a single plane. A "Nichol prism" is a device made up of two pieces of Iceland spar such that one of the two polarized rays passes through and the other is deflected. In a "polarimeter" there are two Nichol prisms through which the light ray passes in turn; the ray will pass with undiminished intensity only if the polarization planes of the two prisms are parallel, it will be diminished in intensity it the two planes are at an acute or obtuse angle and no light will pass through the polarimeter at all if the two planes of polarization are at right angles. Jean Baptiste Biot (like Malus, a Frenchman) in 1813 discovered that a number of substances (such as sugar or turpentine) placed between the two Nichol prisms will change the angle of the prisms at which the light fails to be transmitted. Biot recognized that the substances in question rotate the plane of the polarized light through a certain angle and that the second Nichol prism, if originally set perpendicular to the first in order to produce an obscure field, must be rotated through the same angle to produce an obscure field once again. Thus the phenomenon of optical rotation was discovered. Some substances which produce optical rotation (are "optically active") do so only in the solid (crystalline) state; in such cases the rotation is clearly associated with a property of the crystal. Sir John Herschel, a British astronomer, in 1821 showed that it is the property of dissymmetry in the crystal which is associated with optical rotation. (An object is "dissymmetric" when it is not superimposable with its mirror image--such as a left and right glove; such dissymmetric objects posses a--left or right--handedness.) Other substances, however, produce optical rotation not only in the solid but even in the liquid state (as a melt or solution) or in the vapor phase. In such substances the rotation is clearly associated with the structure of the molecules themselves. Biot's formula for optical rotation $\alpha = [\alpha]_{\lambda}^{t} \cdot l \cdot d$ (appropriately called "Biot's Law")--where α is rotation in angular degrees, l is path length (within the substance or solution) and d is density or concentration--suggests that the rotation is proportional to the number of molecules encountered. (This number also is proportional to the length of the path within the substance and to the density or concentration.) The proportionality constant $[\alpha]_{\lambda}^{t}$ depends not only on the substance but also on the wave-length of light λ, temperature t and solvent. (Solvent and approximate concentration are stated in parenthesis following the formula; concentration may affect $[\alpha]$ by affecting the medium.)

Important discoveries in the area of stereochemistry were made by Louis Pasteur in Paris in the 1850's. Pasteur was aware of the existence of two acids derived from argol, the deposit formed in wine casks; tartaric acid (now called (+)-tartaric acid) which is dextrorotatory and racemic acid (now called (±)-tartaric acid) which does not rotate polarized light. In looking at crystalline salts of tartaric acid (e.g. the sodium ammonium salt) under magnification, Pasteur noticed that the crystals were dissymmetric (see above) and that all had the same handedness. Looking at corresponding crystals of the sodium ammonium salt of racemic acid, he found that they were dissymmetric also but differed in handedness, some having the same handedness (or "chirality") as the salt of tartaric acid, some having the opposite chirality. With the aid of a lens and a pair of tweezers, Pasteur separated the two types of crystals found in the salt of the racemic acid; when he subsequently dissolved them separately in water, he found that the solutions were now optically active, one rotating to the right just as the solution of a salt of tartaric acid and the other one rotating equally to the left. In other words, by a mechanical process Pasteur had brought about what is now called "resolution" of the original mixture of equal numbers of molecules of opposite chirality (the so-called "racemic modification"). Pasteur correctly ascribed the optical activity of the (+)-tartaric acid to a dissymmetry not just of its crystals (for this dissymmetry disappears in solution) but of its very molecules. The unnatural (-)-tartaric acid which he produced has the mirror image configuration and the natural racemic acid or (±)-tartaric acid is a mixture of equal amounts of the (+)- and (-)-acids, devoid of optical acitvity by virtue of a process of compensation. Using a type of projection which will be explained later, the three types of tartaric acid are shown in Fig. 2. (It might be noted here that there is a fourth kind of tartaric acid, so-called <u>meso</u>tartaric acid

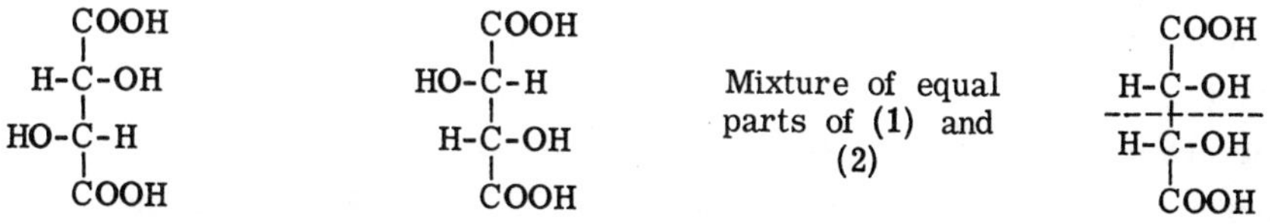

COOH	COOH		COOH
H-C-OH	HO-C-H	Mixture of equal	H-C-OH
HO-C-H	H-C-OH	parts of (1) and	H-C-OH
COOH	COOH	(2)	COOH
(+)-Tartaric acid (1)	(-)-Tartaric acid (2)	(±)-Tartaric acid	Mesotartaric
(<u>R</u>, <u>R</u>)	(<u>S</u>, <u>S</u>)	or racemic acid	acid (3)

Fig. 2

(Formula 3, Fig. 2) which is an inactive stereoisomer of (1) and (2); it has a plane of symmetry (dashed in Fig. 2) and is therefore not dissymmetric ("achiral") and cannot rotate the plane of polarized light under any circumstances. (The reader might note the difference between (±)-tartaric acid--with chiral molecules of opposite handedness present in equal quantities, devoid of rotation by compensation but resolvable into the active (+)- and (-)-forms--and <u>meso</u>tartaric acid with its inherently achiral molecules which are not resolvable.)

The idea that optical activity is due to molecular dissymmetry was further elaborated by Le Bel in France and van't Hoff in Holland in 1874. By this time it had been recognized that carbon is quadrivalent, i.e. tends to link up with four other atoms or groups (Kekulé, 1858). van't Hoff (Nobel Laureate, 1901) suggested that these four atoms or groups were disposed about the central carbon atom in tetrahedral fashion. As has already been seen for the lactic acids (Fig. 1), when the four groups are distinct, two different arrangements are possible which are related to each other as an object and its mirror image. Thus the tetrahedral carbon may serve as a focus of dissymmetry or handedness it is often called an "asymmetric carbon atom" or (in more modern terminology a "chiral center" (Greek cheir = hand). The two different arrangements are the dextrorotatory and levorotatory enantiomers (mirror image isomers) of the chiral molecule; a mixture of equal parts of the two constitutes a racemic modification

III. Basic Concepts and Terminology

a) Symmetry Elements. Whether or not molecules are capable of displaying enantiomerism (i.e. of existing in enantiomeric forms) depends on whether or not they are superimposable with their mirror image. This, in turn, depends on the symmetry point group to which the molecule belongs. Certain symmetry point groups (the so-called C_n and D_n point groups in the Schoenfliess notation) exhibit enantiomerism, other point groups do not. The classification into symmetry point groups is based on the presence or absence of certain symmetry elements, specifically simple axes, alternating axes, centers and planes of symmetry. An n-fold symmetry axis (or axis of order n) is an axis passing through an object (molecule) in such a way that rotation about the axis by an angle of $360^{\circ}/n$ bring the object into a position indistinguishable from its original one. By way of an example, an equilateral triangle has a three-fold axis passing through the center of the triangle at right angles to its plane. The order n can take any positive integral value (e.g. eight for a regular octagon) or it may be infinite (e.g. for a circle). A plane of symmetry is a plane passing through an object such that the part on one side of the plane is the exact reflection of the part on the other side (the plane acting as a mirror). Thus for a book, a plane passing midway between the two covers and bisecting the spine will be a plane of symmetry (if one disregards the printing). A center of symmetry is a point within an object such that a straight line drawn from any part or element of the object to the center and extended an equal distance on the other side encounters an equal part or element. Thus the center of a sphere is a center of symmetry. Finally, an alternating axis of order n is an axis such that when an object containing such an axis is rotated by $360°/n$ about the axis and then reflection is effected across a plane at right angles to the axis, a new object is obtained which is indistinguishable from the original one. This symmetry element is not so easy to visualize but is exemplified by a square wooden board which has two right-handed screws partly drilled into it from one side in the first and third corners and two left-handed screws similarly introduced from the opposite side at the second and fourth corners, perpendicular to the board. (In this particular case the alternating axis is fourfold and passes at right angles to the board through its center.)

It can be shown mathematically that any object having a plane, center or alternating axis of symmetry is superimposable with its mirror image; therefore molecules with these symmetry elements cannot exhibit enantiomerism, i.e. cannot exist in optically active form. In fact, molecules containing each one of these three symmetry elements have been synthesized and have been found to be devoid of optical activity. Such molecules may be called symmetric, but the terms "non-dissymmetric" or "achiral" (i.e. devoid of handedness) are preferred. On the other hand, molecules having only simple axes of symmetry or molecules totally devoid of symmetry elements are not superimposable with their mirror images. Such molecules are "chiral" or "dissymmetric" and exhibit enantiomerism (optical activity). (Clearly the term "asymmetric"--found in the early literature-- is not applicable to such molecules in general, but only to those representatives which are devoid of symmetry axes.)

b) Enantiomerism and Diastereoisomerism. We have already discussed stereoisomers which bear a mirror image relationship to each other; such stereoisomers are called enantiomers (also enantiomorphs, antimers or, less appropriately, optical antipodes). Well-known objects which bear enantiomeric relationships to each other are a right and a left glove, a right-handed and a left-handed screw, a right-handed and a left-handed nut. It is a general proposition that enantiomeric objects bear identical relationships toward symmetric objects.

Thus a right and left glove will fit equally well into a rectangular box; a right and left screw could be equally easily drilled into a wooden board. Moreover, the handedness of a chiral object cannot be established in a completely symmetrical environment, but it can be recognized as being opposite to that of its enantiomer, or it can be correlated with that of another chiral object. For example, the chirality of a right glove may be established as being the same as that of a right hand and opposite to that of a left hand, that of a right-handed screw as being the same as that of a right-handed nut, etc. In fact, it is clear that toward chiral objects, enantiomeric entities do not bear identical relationships (a right glove bears entirely different relationships toward a right and toward a left hand).

What has been said here for macroscopic objects is equally true for chiral molecules. The internal relationship of atoms and groups within a molecule is the same for enantiomers (just as the distance between the thumb and index finger is the same for a right and left glove). As a result, enantiomers behave the same toward achiral chemical reagents or in scalar (non-dissymmetric) physical measurements. They react at the same rate with achiral (symmetric) reagents to give either identical or enantiomeric products. They will also be identical in all scalar physical properties (melting point, vapor pressure, boiling point, refractive index, density, ultraviolet, infrared and nuclear magnetic resonance spectra, mass spectrum, dipole moment, acidity, X-ray and electron diffraction pattern, etc.). They will, however, react at different rates with chiral reagents (as left and right gloves interact differently with a right hand) and they will show different behavior in non-scalar physical measurements, such as optical rotation or optical rotatory dispersion. A very marked case of distinct behavior of enantiomers toward chiral reagents is seen in the biochemical relationship between chiral substrates and enzymes; usually the difference in reactivity between two enantiomers toward the enzyme is so great that in effect only one of the two will enter into reaction. (This may well explain the earlier-mentioned differences in taste, flavor-enhancing properties and antibiotic properties of enantiomers in their interaction with the highly dissymmetric enzyme systems of the human body.)

Stereoisomers which do not bear a mirror-image relationship toward each other are called diastereoisomers. An example is given by mesotartaric acid (Fig. 2) as compared to, say, (+)-tartaric acid. The two compounds have the same constitution but differ in spatial arrangements, thus they are stereoisomers. Not being enantiomers, however, they must be classified as diastereoisomers. Diastereoisomers, in general, have quite different physical properties, for, in contrast to the situation in enantiomers, corresponding atoms and groups in diastereoisomers do not bear the same spatial relationship to each other. For example, if one orients molecular models of the tartaric acids in such a way that the carboxyl groups (which are bulky and experience some dipole repulsion) are as far from each other as possible (see below under "conformation"), then the hydroxyl groups will be close together in (+)-tartaric acid but not in meso-tartaric acid (Fig. 3). (The situation in (-)-tartaric acid is, of course, the same as that in the (+)-enantiomer.) Because of this difference in interatomic

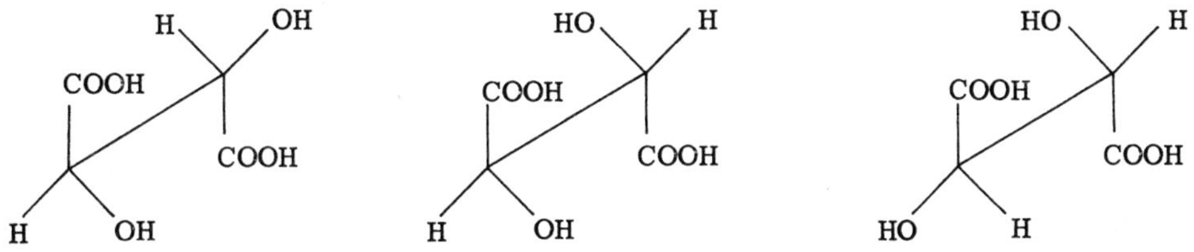

(+)-Tartaric acid   Mesotartaric acid   (-)-Tartaric acid

The tartaric acids (three-dimensional)

Fig. 3

relationships in the molecules, diastereoisomers show different behavior both physically and chemically. For example, (+)-tartaric acid (and its (-)-enantiomer) melts at $174^{\circ}C$ whereas the meso form melts at $151^{\circ}C$; the latter is less dense, less soluble in water and a less strong acid than the former.

Diastereoisomerism will occur in molecules containing more than one chiral center (such as the tartaric acids) as well as in olefins of the type abC=Ccd (where a $\neq$ b and c $\neq$ d) and in cyclic compounds in which two or more ring atoms have unlike substituents; examples are shown in Fig. 4.

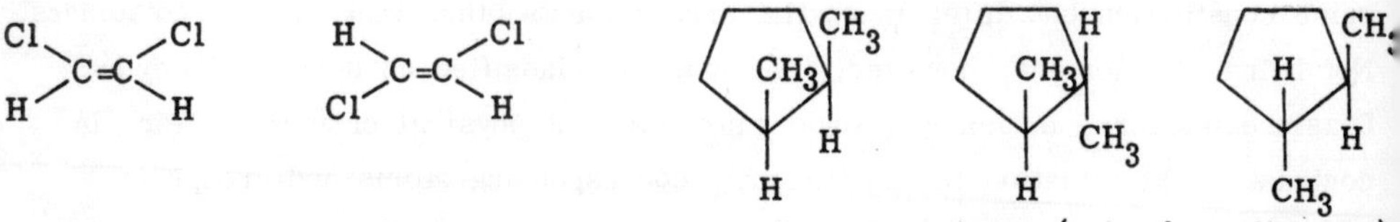

Cl   Cl   H   Cl   CH$_3$ H   CH$_?$

  C=C   C=C   CH$_3$ CH$_3$ H

H   H   Cl   H   H   CH$_3$ CH$_3$ H

                                 H   H

  cis   trans   cis   trans (pair of enantiomers)

1, 2-Dichloroethylenes   1, 2-Dimethylcyclopentanes

Diastereoisomerism in olefins and cyclanes

Fig. 4

The latter type of isomerism is sometimes called cis-trans isomerism or geometrical isomerism to distinguish it from the "optical isomerism" mentioned earlier.
However, the distinction between optical and geometrical isomerism is not a clean-
cut one as is seen in the 1, 2-dimethylcyclopentanes in which the cis isomer is a
meso form (see below) whereas the trans isomer exists as a pair of enantiomers
("dl-pair"); this cyclane system thus presents both diastereoisomerism (as between the cis and either of the trans isomers) and enantiomerism (as between the
two mirror-image trans isomers).

Once it is understood that diastereoisomers differ in chemical and physical
properties, including stability, it may be appreciated why enantiomeric molecules,
while equally reactive toward symmetrical reagents, behave differently in reaction
with other chiral molecules, such as enzymes. For the activated complex which
forms in a reaction of a dissymmetric molecule with a chiral reagent will not be
enantiomeric with the activated complex formed in the reaction of the mirror-
image molecule with the same reagent, but rather the two activated complexes
will be diastereoisomeric and therefore unequal in energy. Thus one enantiomer
passes through the (lower-energy) activated complex faster than the other
enantiomer passes through its (higher-energy) diastereoisomeric activated complex;
the former enantiomer will thus react faster than the latter. An example is provided, once again, by gloves and hands; while a left and right glove react equally
easily with a rectangular box (symmetric reagent) they will not react equally
easily with a left hand (chiral reagent); in this case the left glove forms the much-
lower energy complex (gloved hand) and will therefore react much faster--in fact,
the right glove will not react at all! This analogy explains, for example, why

sarcolactic acid but not fermentation lactic acid is dehydrogenated to pyruvic acid by beef heart lactic dehydrogenase; here the two lactic acids correspond to the two gloves and the chiral enzyme corresponds to the left hand which will accept (and make react) one of the enantiomers but not the other.

c) Projection Formulas. Stereochemical Nomenclature. Molecules are three-dimensional entities and are therefore difficult to represent on two-dimensional paper, unless appropriate modes of projection are agreed on. There are four relatively standard such modes, illustrated in Fig. 5 for a molecule of the constitution $CH_3CHBrCHOHCH_3$. (This molecule may exist in four stereoisomeric forms, as will be explained in the next section. Only one form is shown in the four different systems in the first row of Fig. 5; all four forms are shown in the Fischer projection in the second row.) In the "flying-wedge" projection, the molecule is seen side-on, in its normal "conformation" (see below) or

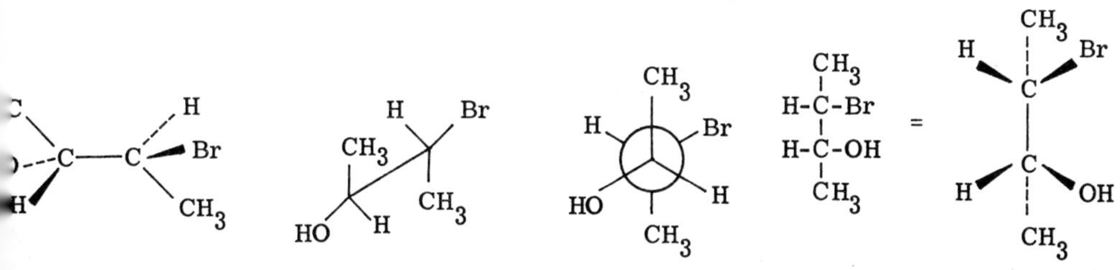

"Flying-wedge   "Saw-horse   "Newman   "Fischer projection"

Enantiomers   Enantiomers
(threo Isomer)   (erythro Isomer)

Diastereoisomers

Two-dimensional representations of 3-bromo-2-butanol

Fig. 5

rotational arrangement of the groups, which corresponds to a "staggered" arrangement. In the "saw-horse" projection, which gives a better view, the molecule is seen at an angle. In the "Newman projection" the molecule is seen front-on. The Fischer projection, in contrast to the three others, captures the molecule in an eclipsed conformation with the C_2-C_3 axis in front, the C_1 and C_4 CH_3-groups pointing backward and the H, OH and Br substituents at C_2 and C_3 pointing forward. Simpler Fischer projection formulas for molecules containing a single chiral center are shown in Fig. 6; these correspond to sarcolactic acid (or (+)-lactic acid) and fermentation lactic acid ((-)-lactic acid) represented in Fig. 1. (The tetrahedra shown in Fig. 1 must be turned 180°

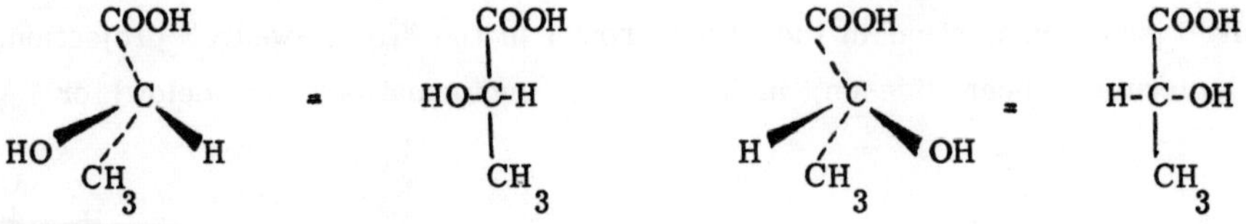

(+)-Lactic acid
(S)

(-)-Lactic acid
(R)

Fig. 6

around a vertical axis to be in the position shown in Fig. 6.) The top and bottom groups (CH_3, COOH) by convention are behind the projection plane and the sideways groups (H, OH) in front of this plane.

In addition to a convenient two-dimensional representation of a chiral center one needs a way of including the arrangements of the groups at such a center ("configuration") in the standard name of the compound. This is now achieved through the Cahn-Ingold-Prelog system [see Angew. Chem., Int. Ed., 5, 385 (1966)] by assigning a prefix (R)- or (S)- to each chiral center, depending on its configuration. In order to assign the appropriate prefix at a chiral center Cabcd, one arranges the groups a, b, c and d according to a sequence rule (see below); if the sequence is a ← b ← c ← d, one then views the chiral center from the side opposite the group of lowest rank (d): if the remaining groups a, b, c then occur in clockwise order, the prefix is (R)-, if they occur in counterclockwise order, the prefix is (S)- (Fig. 7).

R-S Nomenclature

Fig. 7

It is easy to see with models that, in a Fischer projection formula, if atom d (lowest sequence) is placed at the bottom, then if $a \rightarrow b \rightarrow c$ describes a clockwise array, the prefix is (R), if it describes a counterclockwise array, (S). If a given Fischer projection has a group other than that of lowest sequence at the bottom, it may be properley adjusted by exchanging groups in three's or by effecting two exchanges of two groups each (a single such exchange will invert the configuration). To obtain the sequence a,b,c,d, the following "sequence rules" are used: 1) atoms attached to the chiral center are ordered according to atomic number, e.g. $Br > Cl > C > H$. 2) if the atoms attached to the chiral center are alike, then, for the like atoms only, one goes on to the next atom out and, if necessary, to the next one after that, etc. Under this rule, $CH_3-CH_2 > CH_3$ (because $C > H$) and $CH_3-CH_2-CH_2 > CH_3-CH_2$ (same reasoning, one atom further out). If two substituents have atoms of the same rank attached, but in unequal numbers, the substituent with more atoms of the highest rank takes precedence. This leads to the sequence $(CH_3)_3C > (CH_3)_2CH > CH_3-CH_2$ (3 C's > 2 C's > 1 C) and $CH(OR)_2 > CH_2OR$ (2 O's > 1 O). 3) multiply bonded atoms are complemented by duplicating (or, in the case of triple bonds, triplicating) the ligands at both ends of the bond; for atoms having less than four ligands otherwise, the missing ligands are replaced by hypothetical ligands of atomic number zero. Under this rule CHO becomes

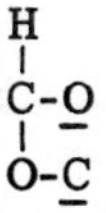

H
|
C-O   (duplicated ligands underlined) and thus has precedence over CH_2OH (but
|
O-C

not over $CH(OCH_3)_2$), the ligand carbon in C_6H_5 becomes

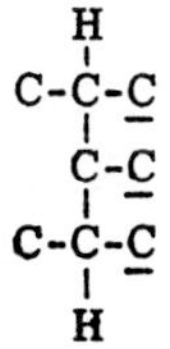

```
      H
      |
    C-C-C
      |
      C-C
      |
  C-C-C
      |
      H
```

and has

precedence over $C(CH_3)_3$, and $NH(CH_3)_2^+$ has precedence over $N(CH_3)_2$ (which
becomes $\overset{..}{N}(CH_3)_2$). 4) when the only difference is due to isotopic substitution,
the atom of higher mass number takes precedence (e.g. D > H). When the only
difference between two ligands is in configuration, seqcis > seqtrans and R > S
(see below). By way of application of these rules, proper configurational symbols
are included in Figs. 2, 5 and 6. It should be noted that when there are several
chiral centers (Figs. 2, 5), a symbol is attached to each; thus the first of the
four stereoisomers of 3-bromo-2-butanol shown in Fig. 5 (bottom part) is (2S, 3S)-
3-bromo-2-butanol and the (+)-tartaric acid formula shown in Fig. 2 is (2R, 3R)-
tartaric acid.

Other systems of configurational nomenclature (using prefixes such as D,
L or α, β) are found in the literature but are less general and frequently less
easy to apply than the Cahn-Ingold-Prelog system.

d) Number of Stereoisomers. Meso Forms. Pseudoasymmetric Atoms.
Since a chiral atom has two possible arrangements (configurations), for each
chiral atom in a molecule the number of stereoisomers is normally doubled.
Thus, if there are n chiral atoms, the number of stereoisomers will be 2^n.
Since normally for each configurational arrangement there exists an enantiomeric
(mirror-image) arrangement, the 2^n stereoisomers group themselves in 2^{n-1}
diastereoisomeric pairs of enantiomers. However, there are "degenerate cases"
where some of the purported isomers are, in fact, indistinguishable and the total
number of isomers is less than the formula suggests. This will happen when two
or more of the chiral centers in the molecule are alike, as, for example, in the
tartaric acids (Fig. 2) or the 1, 2-dimethylcyclopentanes (Fig. 4). The tartaric
acids have two chiral centers (*CHOHCOOH) which are alike; the total number of
isomers is three; the (+) and (-)-forms and the inactive stereoisomer. (The
racemic modification or dl-pair being a mixture is not counted separately.) The
stereoisomerism of the 1, 2-dimethylcyclopentanes is exactly analogous. The
inactive diastereoisomer in these sets is called the "meso form", such as meso-
tartaric acid in Fig. 2 and meso-1, 2-dimethylcyclopentane (alternatively called
cis-1, 2-dimethylcyclopentane) in Fig. 4. (Usually the term meso is not applied to
sets of stereoisomers which do not contain optically active forms, such as the
1, 2-dichloroethylenes in Fig. 4.)

In certain molecules one finds atoms which, through a change in configuration, give rise, not to a pair of enantiomers, but to a pair of diastereoisomers. Such atoms are called "pseudoasymmetric". They are characterized by the fact that, of their four ligands, two are distinct (but achiral) and the other two are enantiomeric, as expressed in the formula $CXYZ_R Z_S$.

The trihydroxyglutaric acids shown in Fig. 8 constitute a set in which pseudoasymmetry occurs. In the first two isomers, the central (No. 3) carbon atom is clearly achiral (notwithstanding the fact that the molecules as a whole are chiral), since the attached groups (-CHOHCOOH) are identical within each

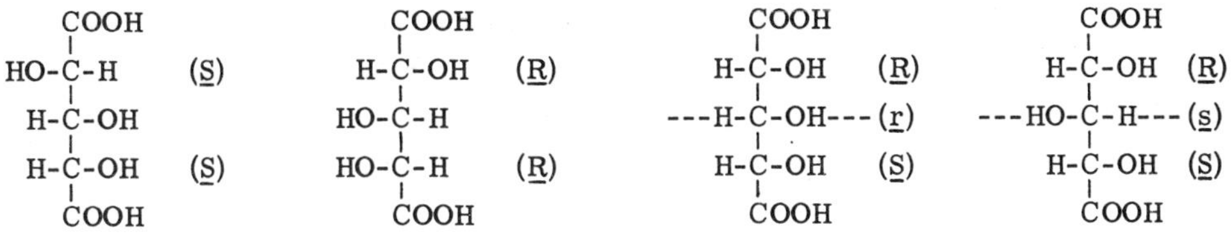

COOH		COOH		COOH		COOH	
HO-C-H	(S)	H-C-OH	(R)	H-C-OH	(R)	H-C-OH	(R)
H-C-OH		HO-C-H		---H-C-OH---	(r)	---HO-C-H---	(s)
H-C-OH	(S)	HO-C-H	(R)	H-C-OH	(S)	H-C-OH	(S)
COOH		COOH		COOH		COOH	

(-)	(+)	meso-1	meso-2
Enantiomers		meso Forms	

The trihydroxyglutaric acids

Fig. 8

isomer. Here changing the arrangement about the No. 3 carbon leaves the molecule unaltered (the changed arrangement may be converted to the original by turning the molecule upside down). In the remaining two stereoisomers (which are achiral meso forms), however, the No. 3 carbon atom fulfills the conditions of being pseudoasymmetric: a change in its configuration changes one achiral diastereoisomer (meso-1) into another (meso-2). The configurational symbols at pseudoasymmetric atoms are given as r and s (instead of R and S); in other respects the configurational symbols in Fig. 8 correspond to the rules given earlier.

In molecules of the type ABxC-CABy in which the two chiral centers are dissimilar, there are, of course, two diastereoisomeric pairs of enantiomers (Fig. 5). One of these has a formal resemblance to the active isomers in the corresponding molecule ABx-ABx and the other has a formal resemblance to the meso isomer. The former pair is called "threo" and the latter "erythro", as shown in Fig. 5. This type of short common nomenclature is useful because pairs of threo- and erythro-isomers are often used in stereochemical investigations. The names are derived from the names of the four-carbon sugars threose and erythrose, $CH_2OHCHOHCHOHCHO$ having the appropriate arrangements; in fact, much of the early interest in stereoisomerism was derived from the chemistry of sugars, many of which correspond to the general formula $CH_2OH(CHOH)_nCHO$ and have n chiral centers.

e) Racemic Modifications. Resolution. A "racemic modification" has already been defined as an assembly of equal numbers of enantiomeric molecules, i.e. dextrorotatory (+) and levorotatory (-) molecules of the same substance. Clearly, racemic modifications exist only at the macroscopic, not at the molecular level; individual molecules, if chiral at all, are either left-handed or right-handed but not both. Depending on their phase behavior, racemic modifications may be either "racemic mixtures", "racemic compounds" (also called "racemates") or "racemic solid solutions". None of these terms should therefore be used synonymously with "racemic modification". It might be noted that a racemic modification (or "racemic form") with equal numbers of two types of molecules corresponds to a more random or disorderly state of affairs than a pure enantiomer (one single type of molecule), therefore there is an increase of randomness and therefore an increase in entropy as one passes from one enantiomer or other to the racemic modification. This process called "racemization" is thus thermodynamically favorable and may be brought about in a number of ways (e.g. by heat, or by treatment with acid, base, metal catalysts); it cannot, unfortunately, be dealt with here in a general way, for the question whether racemization can be effected for a given enantiomer and, if so, how, depends on the structure and detailed chemical behavior of the substance. It should be pointed out, however, that synthesis (in the absence of chiral reagents) of chiral compounds from achiral precursors always leads to racemic modifications. The point is illustrated in Fig. 9 for the two most general ways of creating chiral centers: by substitution and by addition. The example for substitution is synthesis of 2-bromopropanoic acid by bromination of propanoic acid; since either of the two hydrogen atoms of propanoic acid is equally easily replaced by bromine, the two enantiomers of 2-bromopropanoic acid must necessarily be formed in equal amounts

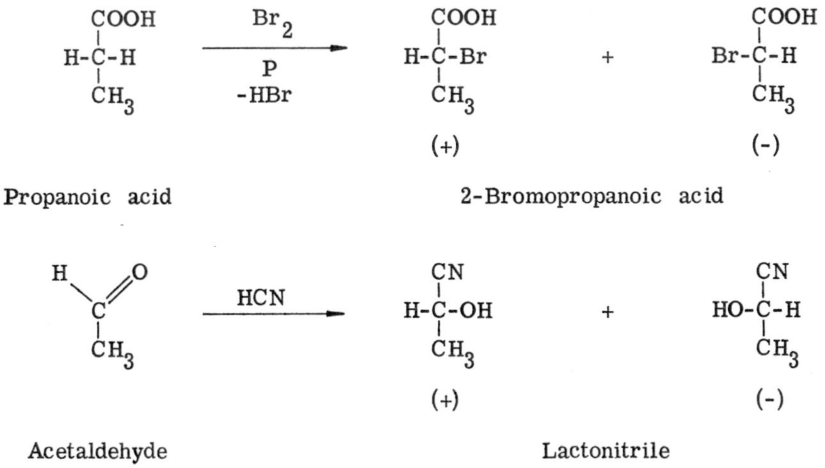

$$\underset{\text{Propanoic acid}}{\overset{\displaystyle COOH}{\underset{\displaystyle CH_3}{H-\overset{|}{\underset{|}{C}}-H}}} \quad \xrightarrow[\substack{P \\ -HBr}]{Br_2} \quad \underset{(+)}{\overset{\displaystyle COOH}{\underset{\displaystyle CH_3}{H-\overset{|}{\underset{|}{C}}-Br}}} \quad + \quad \underset{(-)}{\overset{\displaystyle COOH}{\underset{\displaystyle CH_3}{Br-\overset{|}{\underset{|}{C}}-H}}}$$

2-Bromopropanoic acid

$$\underset{\text{Acetaldehyde}}{\overset{\displaystyle H}{\underset{\displaystyle CH_3}{C}}\diagdown\negmedspace\diagup O} \quad \xrightarrow{HCN} \quad \underset{(+)}{\overset{\displaystyle CN}{\underset{\displaystyle CH_3}{H-\overset{|}{\underset{|}{C}}-OH}}} \quad + \quad \underset{(-)}{\overset{\displaystyle CN}{\underset{\displaystyle CH_3}{HO-\overset{|}{\underset{|}{C}}-H}}}$$

Lactonitrile

Racemic modifications by synthesis

Fig. 9

and a racemic modification results. Similarly, in the reaction of hydrogen cyanide with acetaldehyde to give lactonitrile, the two sides of the C=O double bond are equally easily approached by cyanide and again equal amounts of the two enantiomeric lactonitriles must result.

Since ordinary synthesis leads to racemic modifications, two questions immediately arise, namely how are pure enantiomers obtained in the laboratory and why is it that many substances occurring in nature--such as the lactic acids, (+)-tartaric acid, sugars such as glucose, alkaloids such as quinine, steroids such as cholesterol, vitamins such as ascorbic acid (Vitamin C), polypeptides and proteins such as insulin, nucleic acids such as DNA, antibiotics such as penicillin and many others--occur in a single, enantiomerically pure form? (A few substances, such as the lactic acids, actually occur as either enantiomer, but the two enantiomers are found in different places and do not normally occur together as a racemic modification.)

As far as obtaining pure enantiomers in the laboratory is concerned, several methods are available, but all of them, except possibly the first, require the availability of other pure enantiomers from natural sources. The first method is the method of separating enantiomers by mechanical separation of enantio-

morphous crystals, as in the resolution of tartaric acid originally effected by
Pasteur and described earlier. (By "resolution" is meant the separation of the
enantiomers in a racemic modification.) This method is the only one which doe
not require the availability of another chiral substance (see below) but, besides
lacking generality (only in racemic mixtures, but not in compounds or solid
solutions do the enantiomers crystallize separately), it is tedious even when
applicable because of the problem of growing large crystals and the difficulty
or sometimes virtual impossibility of distinguishing crystals of opposite handed
ness. Other methods are therefore generally employed of which the two most
important are resolution via diastereoisomeric salts and resolution by enzymat:
processes.

In resolution via diastereoisomeric salts, the compound to be resolved is
usually first converted, by appropriate chemical processes, into an acid or bas
(If it is already an acid or base, this preliminary step is unnecessary.) Suppos
then, that the compound to be resolved is an acid (±)-A. It is then combined,
solution, with an optically active base, say (-)-B, usually (but not necessarily)
of natural origin--for example brucine. Two salts are formed, (+)-A.(-)-B an
(-)-A.(-)-B. (It must be remembered that, even though (±)-A is a racemic
modification, it consists of individual (+)-A and (-)-A molecules.) Since the tw
salts are diastereoisomeric, they will, in general, be different in physical
behavior (see above); for example, they will differ in solubility and one will
crystallize in preference to the other. The less soluble salt (say (+)-A.(-)-B) i
then collected, recrystallized until pure, and finally decomposed with strong aci
to give the pure enantiomer (+)-A. (The "resolving agent" (-)-B may generally
by recovered and used again.) The more soluble salt, (-)-A.(-)-B may sometimes be crystallized until pure also, but more generally impure (-)-A is
recovered from it by acid treatment and purified by conversion to another, les
soluble salt, e.g. (-)-A.(+)-B (if (+)-B is available--in the case of brucine it i
not) or (-)-A.(-)-C (C is another optically active base, for example cinchonidir
which is then crystallized to purity and finally decomposed to give pure (-)-A.

In the enzymatic method, various approaches are possible. They are
generally based on the fact that the enzyme systems of living organisms are
attuned to that enantiomer which is of normal natural occurrence. For exampl
Pasteur discovered a mould, <u>Penicillium glaucum</u>, which is able to metabolize

(and thus destroy) the naturally-occurring (+)-tartaric acid but not the unnatural (-)-tartaric acid. When synthetic (±)-tartaric acid is inocculated with this mould, the (+)-molecules are metabolized and destroyed and the (-)-molecules remain behind; isolation of the residual tartaric acid from the fermentation mixture thus gives pure (-)-tartaric acid. (The reason for the discrimination of the chiral enzyme for one of two enantiomeric substrates has been explained earlier.) In another version of the process, one of the two enantiomers, instead of being destroyed, is chemically transformed. For example, in the resolution of synthetic (±)-amino acids to give the active amino acids (which are the building blocks of all natural proteins), the racemic amino acid is first acetylated to give a racemic acetylamino acid. The acetylamino acid is then hydrolyzed in the presence of the enzyme hog kidney acylase (the raw material for the isolation of this enzyme is obtained from a slaughter house). The enzymatic hydrolysis affects only the acetyl derivatives of the natural (generally S̲) amino acids which are thus obtained in the free state, readily isolable from the residual acetyl derivatives of the (R̲)-acids. The free (R̲)-amino acids may be obtained by hydrolyzing the (R̲)-acetyl-amino acids by ordinary chemical means (e.g. in the presence of hydrochloric acid).

Only the major means of resolution have been described here. Other means--resolution by various forms of chromatography, crystallization from optically active solvents, asymmetric transformation or asymmetric synthesis under conditions other than enzymatic ones, etc. are principally of theoretical interest.

One of the interesting aspects of resolution is that it generally involves a propagation rather than a creation of dissymmetry. It is true that a relatively small amount of a resolving agent (e.g. an optically active base) or of an enzyme can serve to resolve a relatively large amount of racemic material given enough time. There remains, however, the question of the ultimate origin of dissymmetric materials in nature. Essentially two possibilities have been discussed. One is a process rather akin to Pasteur's mechanical separation of enantiomers, except that it involves the nucleation of a saturated solution of a racemic material with a seed crystal of one of the two enantiomers or by a pseudomorphic crystal (similarly shaped crystal of a different substance). A seed crystal could have formed accidentally and spontaneously (as it does in other spontaneous crystallizations) or a pseudomorphic dissymmetric crystal could have been obtained from a substance, such as urea, which is not dissymmetric on the molecular scale but is dissymmetric in the crystalline state. If, after seeding with one enantiomer the mother liquor is accidentally separated (e.g. by spilling) from the enantiomeri crystals, resolution occurs. The other possibility of the origin of optically active material is by asymmetric synthesis in the presence of a dissymmetric physical rather than chemical agent—such as circularly polarized light formed, for example, by reflection of ordinary light, e.g. from a lake. Photochemical reactions occurring in the presence of circularly polarized light have been shown to lead to optically active compounds.

The efficacy of resolution is measured by the "optical purity" of the resolved material, meaning the ratio of the rotation of the material obtained to the maximu rotation possible for the pure enantiomer of the same material.

f) Enantiotopic and Diastereotopic Groups. When two atoms or groups in a molecule may be superimposed by a rotation of the molecule such that the new arrangement is indistinguishable from the old, the groups are considered equivalent. An example is provided by the hydrogen atoms in dichloromethane, CH_2Cl_2. The implication of the above statement is that the molecule has a symmetry axis which is therefore a necessary and sufficient condition for the existence of equivalent atoms or groups. Replacement of one or other of two (or more) equivalent groups by a new atom or group gives a single new molecule; thus replacement of either hydrogen in CH_2Cl_2 by bromine gives the same dichlorobromomethane, $CHBrCl_2$.

In contrast, there are atoms or groups which appear to be alike but which are not superimposable by a process of rotation. The two hydrogens on the CH_2-group in propanoic acid (Fig. 9) are of this type. It might be noted that replacement of one or other of these hydrogens by bromine (Fig. 9) does not give rise to the same molecule but rather two mirror image molecules—the R and S forms of 2-bromopropanoic acid—result. Clearly the two designated hydrogens in propanoic acid are not equivalent but rather bear a mirror-image relationship to each other, as do the bromine atoms in the (R)- and (S)-forms of 2-bromopropanoic acid. In the latter case, there are two different molecules which are called enantiomeric. In propanoic acid, the mirror-image relationship is between atoms within the same molecule; since these atoms thus differ in place (Greek "topos"), they are called "enantiotopic". The most important property of enantiotopic atoms or groups is that they behave differently in a dissymmetric environment--for example in the presence of an enzyme. Thus it has been shown through labeling experiments that only one of the two enantiotopic hydrogens in ethanol, $CH_3-\overset{\displaystyle H}{\underset{\displaystyle H}{C}}-OH$ is removed in the oxidation of the molecule to acetaldehyde in the presence of the enzyme alcohol dehydrogenase. A classical case of the importance of enantiotopic groups is provided through a consideration of the metabolic path of the conversion of oxaloacetic acid, $HOOCCOCH_2COOH$ and acetic acid, CH_3COOH to α-ketoglutaric acid, $HOOCCOCH_2CH_2COOH$ by pigeon breast enzymes. When labeled oxaloacetic acid, $HOOCCOCH_2COOH^*$ was the precursor, only the carboxyl group next to the carbonyl was labeled in the product, $^*HOOCCOCH_2CH_2COOH$. It had previously been thought that the first step in the biochemical transformation was an aldol-type addition of acetic acid (or acetate) to oxaloacetic acid to give citric acid, $HOOCCOH(CH_2COOH)_2$ which was then decarboxylated, dehydrated, rehydrated and oxidized to α-ketoglutaric acid. However, it appeared that since the two $-CH_2COOH$ groups in citric acid were the same, such a reaction path could not possibly lead to an α-ketoglutaric acid labeled at one end only. It is now clear that the argument against the intermediacy of citric acid is not cogent, however. The two $-CH_2COOH$ groups of citric acid are not, in fact, equivalent but are enantiotopic. Since the citric acid is formed in the presence of an enzyme or enzymes, one of the enantiotopic groups originates cleanly from the oxaloacetic acid (and is therefore labeled in the COOH) whereas the other originates cleanly from the acetic acid (and is therefore not labeled in the COOH). In the further transformation of the citric acid by a different enzyme the two enantiotopic groups are again distinct and it happens that the group derived from the oxaloacetic acid (the labeled one) is the one which ends up next to the CO group of the α-ketoglutaric acid. This has nothing to do with the label as such, but rather with the spatial disposition of the two enantiotopic CH_2COOH groups in the citric acid intermediate, as may be shown easily by synthesizing the citric acid from unlabeled $HOOCCOCH_2COOH$ and labeled CH_3COOH^*. In this synthesis the label is placed in the other enantiotopic CH_2COOH-group of the citric acid and it now ends up remote from the CO-group in the product, $HOOCCOCH_2CH_2COOH^*$. A careful analysis of the citric acid synthesis reveals as an important feature the fact that the two faces of the carbonyl ($C=O$) group in oxaloacetic acid are themselves not equivalent but are enantiotopic; enzymatic synthesis of citric acid entails exclusive approach of CH_3COOH from one side to this carbonyl group. By the same reasoning, reduction of acetaldehyde, $CH_3\overset{H_a}{\underset{}{C}}=O$ in the presence of alcohol dehydrogenase leads to ethyl alcohol, $CH_3\overset{H_a}{\underset{H_b}{C}}-OH$ in which the hydrogen originally present and that introduced are stereochemically distinct; if the material reduced is labeled, i.e. $CH_3\overset{D}{\underset{}{C}}=O$, the $CH_3\overset{D}{\underset{H}{C}}-OH$ product will be a single, optically active enantiomer.

Groups or atoms whose replacement gives rise to diastereoisomers are neither equivalent nor in mirror image location but are truly distinct in any environment, chiral or achiral. Such groups are called "diastereotopic". An example is provided by the methylene hydrogens in 2-butanol, $CH_3\overset{H}{\underset{H}{C}}CHOHCH_3$. As may be appreciated from Fig. 5 (bottom row, second and third formula), replacement of one or other hydrogen in a given enantiomer of 2-butanol by bromine gives rise to diastereoisomeric 3-bromo-2-butanols which, being different in energy content, are formed in unequal proportions. Thus the diastereotopic hydrogen atoms, in addition to giving rise to diastereoisomeric products, are unequally reactive even toward achiral reagents. In fact, they find themselves in an entirely different chemical environment (not just in a mirror-image environment, as enantiotopic atoms) and are distinct physically as well as chemically. For example, in nuclear magnetic resonance spectroscopy the two hydrogens can be discerned by their distinct resonance signals.

g) Configuration. We have already seen that by "configuration" of a molecule is meant the distinctive arrangement of atoms or groups about the chiral part or parts of the molecule (the chiral center in the cases so far discusse but see below for other chiral elements). Thus lactic acid may exist in the (R)- and (S)-configurations shown in Fig. 6. The question, of course, arises as to which of the two experimentally known lactic acids, the dextrorotatory and the levorotatory isomer, corresponds to which spatial arrangement of the groups; in other words "is the configuration of (+)-lactic acid ($\underline{R}$) or ($\underline{S}$) ? " This question was not actually answered until 1951 when Bijvoet, Peerdeman and van Bommel (working in the same laboratory occupied 77 years earlier by vant'Hoff) by a special technique called X-ray fluorescence (not ordinary X-ray diffraction which is incapable of distinguishing mirror images) found that (+)-tartaric acid has the configuration shown in Fig. 2, i.e. ($\underline{R}, \underline{R}$). The configuration of most other compounds has been determined by correlating them directly or indirectly with (+)-tartaric acid using the principle of minimum structural change. Thus the configuration of (-)-lactic acid (Fig. 6) has been determined by chemically replacing one of the CHOHCOOH groups in (+)-tartaric acid (Fig. 2) by a CH_3 group (this requires several successive chemical operations); since (+)-tartaric acid in this transformation is converted to (-)-lactic acid, the configuration of the latter is as shown in Fig. 6, i.e. ($\underline{R}$). [The coincidence of the configurational symbols of (+)-tartaric and (-)-lactic acids--both ($\underline{R}$)--is accidental and no significance should be attributed to it. One of the intermediates in the correlation, (-)-bromolactic acid (CH_2Br instead of CH_3 in the formula for (-)-lactic acid in Fig. 6) has the symbol ($\underline{S}$) although it has the same configuration, clearly, as the (-)-lactic acid to which it is reduced.]

Several ingenious methods of determination of configuration other than chemical correlation have been developed, such as the method of quasi-racemates, mechanistic correlations and methods depending on optical rotatory dispersion measurements. These methods are beyond the scope of this article but may be found in the monograph on stereochemistry (Eliel, 1962) cited at the end. Clearly, since configuration (like constitution) is an essential ingredient of structure, its determination is important; accurate predictions of the physical, chemical and biochemical behavior of chiral compounds cannot be made until the configuration of the compounds is established experimentally.

h) Conformation. Conformational Analysis. A molecule such as ethane (Fig. 10) may, in principle, exist in an infinite number of arrangements differing by the positions of the hydrogen atoms on C_1 relative to those on C_2. Clearly these arrangements, called "conformations", are interconverted by rotation about the C-C single bond. Until the 1930's it was believed that rotation about single bonds was essentially free (i.e. immeasurably fast) so that individual arrangements of the type shown in Fig. 10 could not be detected. However, in the early 1930's

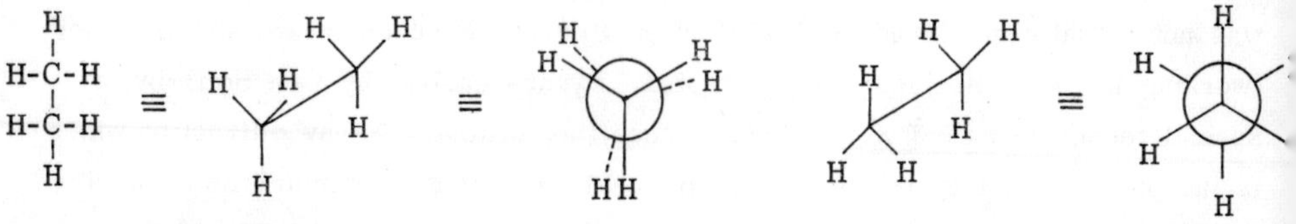

"Eclipsed"   "Staggered"

(intermediate conformations are also possible)

Ethane

Fig. 10

evidence was obtained that this point of view was erroneous and in 1936 Pitzer
showed that the experimental and calculated thermodynamic properties of ethane
can be reconciled only if one assumes a barrier of approximately 3.0 kcal./mole
to rotation about the carbon-carbon bond. Subsequent investigation showed that
the energy maximum occurs when the hydrogens on different carbon atoms are
as close to each other as they can be ("eclipsed conformation", see Fig. 10)
and the minimum corresponds to maximum distance of the hydrogens ("staggered
conformation", Fig. 10). It might be thought that the preference for the staggered
conformation results from a spatial or steric interaction of the hydrogen atoms
("steric hindrance"), but calculation shows that this is not so; the size of the
hydrogen atoms is insufficient to produce a steric interference of more than about
0.3 kcal./mole in the eclipsed position. The remainder of the barrier energy
must be of different origin; present indications are that it is due to a quantum
mechanical repulsion of the electrons in the bonds (R. Pitzer, 1968).

As a result of the barrier, most molecules of ethane are at or near the
staggered conformation at room temperature; there will be a slight torsional
oscillation ("libration") of the molecules to and fro about this energy minimum.
In a molecule such as 1,2-dibromoethane (Fig. 11) the conformational situation
is more complicated; there will now be three energy minima (two of them
corresponding to enantiomeric conformations and therefore equal in energy, the
third one different) and three maxima (two equal, the third different). The
conformations and their names and energies are depicted in Fig. 11.

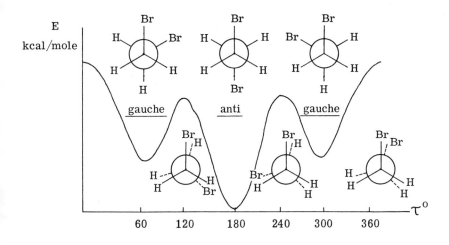

Conformations and Energetics of 1,2-Dibromoethane

Fig. 11

The horizontal axis (abscissa) measures the so-called "torsional angle" τ, defined in this case as the angle between the $Br-C_1-C_2$ and C_1-C_2-Br planes. The vertical axis (ordinate) measures energy. Energy maxima occur at 0 and 120° (with repetitions at 240 and 360°) corresponding to eclipsed conformations; the Br-Br eclipsed conformation is the one highest in energy because it involves both steric and dipolar repulsions of the bromines. Of the minima, the one at 180° corresponding to the so-called "anti" or "anti-periplanar" form (Greek "anti" = against, opposite; periplanar meaning about planar) is the one of lowest energy because both steric and dipolar interactions are at a minimum. The 60° and 300° minima are called "gauche" (French "gauche" = left or skew), "skew" or "syn-clinal" (Greek "syn" = together, "clinal" meaning inclined) and are higher in energy because of residual steric and dipolar interactions between the bromine atoms. Clearly most of the molecules, at any one time, will be in the anti conformation (which has zero dipole moment because the C-Br bonds are antiparallel), but a sufficient number will be in the gauche conformation (which has a dipole moment) to impart an overall dipole moment to 1,2-dibromoethane. In the gas phase, the energy difference between the gauche and anti forms of

1,2-dibromoethane is about 1.4 kcal./mole; taking into account the Boltzmann distribution of the molecules among unequal energy levels as well as the fact that the gauche conformation has a statistical advantage of 2 (there are two mirror-image gauche forms but there is only one anti form), this leads to a distribution, at room temperature, of about 85% of the molecules in the anti conformation, 15% in the gauche. In the liquid phase the dielectric constant is higher and therefore (by Coulomb's law) the repulsion of the electric dipoles less; therefore the gauche conformation exceeds the anti by only about 0.7 kcal./mole in the liquid phase and about one molecule out of three will be gauche in the liquid phase. The dipole moment of 1,2-dibromoethane may be calculated from the conformational distribution of the molecules and vice versa.

Clearly our knowledge of the molecular architecture of 1,2-dibromoethane is not complete until we know its conformation (or distribution over various conformations at a given temperature). "Structure" (i.e. molecular architecture) therefore comprises conformation (rotational arrangement about single bonds) as well as configuration and constitution. Only with a very simple molecule, such as methane, CH_4, does constitution alone completely and unequivocally define structure. In the slightly more complex case of clorobromoiodomethane, CHClBr structure is not completely defined until configuration as well as constitution is known. In the above case of 1,2-dibromoethane, conformation as well as constitution must be known to establish structure. Finally, and most generally, in a compound such as 2,3-dibromobutane, $CH_3CHBrCHBrCH_3$ which exists as meso, (+)- and (-)-stereoisomers, one must know constitution, configuration and conformation before one has the complete information on structure.

By "conformational analysis" is meant an analysis of the physical and chemical properties of a compound in terms of its preferred conformation (or in terms of its conformational distribution if there is more than one preferred conformation, as in $BrCH_2CH_2Br$). Often it is possible to deduce conformation (or conformational distribution) from a few physical or chemical properties and then use the knowledge so gained to predict other physical and chemical behavior (Such prediction, of course, is the aim of structural study.)

Conformational considerations are particularly important in saturated or near-saturated six-membered rings whose properties are strongly dependent on the fact that the rings are chair-shaped (Sachse, 1890). For a detailed treatment of this subject, the monographs on conformational analysis cited at the end of this book should be consulted.

IV. Stereochemistry of Carbon Compounds

Of the various elements whose stereochemistry has been studied, carbon is by far the most important, not just for historical reasons (see above) but because of the universal presence of carbon in the constituents of living matter. We will therefore discuss the stereochemistry of carbon compounds in some detail and then deal briefly with the stereochemistry of other elements.

a) Saturated Compounds. Kekulé, in 1858, realized that carbon is quadrivalent. Since there is no evidence whatsoever that the four hydrogens in the simplest hydrocarbon, methane (CH_4), are in any way distinct or distinguishable, some regular arrangement of the four valencies suggests itself. Three such arrangements come to mind: the square (with the carbon at the center), the square pyramid (with the carbon at the apex) and the tetrahedron (with the carbon at the center). The salient fact which led to a decision between these three possibilities is that there is one and only one isomer of dichloromethane, CH_2Cl_2 (or, for that matter, of any simple molecule of the type Caabb). One may easily convince oneself with a piece of paper and a pencil that a square arrangement for CH_2Cl_2 should lead to two isomers (one with the chlorines adjacent, the other with the chlorines diagonally opposite) and a square pyramid arrangement should similarly lead to two isomers. The tetrahedral arrangement is uniquely compatible with the experimental fact that only one compound CH_2Cl_2 exists. It does, also, predict the existence of enantiomers in the case of Cabcd and was therefore the model adopted by van't Hoff. Experiments carried out since 1874--X-ray diffraction, electron diffraction, neutron diffraction, infrared studies, etc. --as well as quantum mechanical calculations fully bear out van't Hoff's hypothesis; symmetrically substituted molecules, such as CCl_4 (carbon tetrachloride) have the shape of a regular tetrahedron with angles of $109°28'$. [When the substitution is not symmetrical, as, for example, in CH_2Cl_2, the angles will differ slightly from the tetrahedral (∢ Cl-C-Cl, $111.8°$, ∢ H-C-H, $112.0°$, ∢ H-C-Cl, $108.3°$). The

bond distances, C-H and C-Cl are, of course, quite unequal also--C-Cl, 1.77 Å,
C-H, 1.07 Å and the model as a whole deviates quite appreciably from a regular
tetrahedron.]

The stereochemistry of saturated molecules has been discussed adequately
in the preceding section.

b) Olefins. Olefins of the type Cab=Cab exist as stereoisomers (Fig. 4)
the necessary and sufficient condition for this type of isomerism being that a $\neq$ b.
Since all the atoms in Cab=Cab are ordinarily in a plane, this plane is
necessarily a plane of symmetry (unless a or b themselves are dissymmetric
groups) and olefins therefore do not ordinarily exhibit enantiomerism but the
cis (Latin cis = on this side) and trans (Latin trans = across) isomers (Fig. 4)
are diastereoisomers, sometimes called "geometrical isomers". No more than
two isomers are possible in the more general structure Cab=Ccd, but when there
are n non-cumulated (see below) double bonds, the number of diastereoisomers
will be 2^n, except in degenerate cases, such as $CH_3CH=CHCH=CHCH_3$ where,
because of identical substitution of the two double bonds, the number of stereoisomers is reduced (cf. the case of the tartaric acids, Fig. 2). In the case of
2,4-hexadiene shown above, there are three diastereoisomers, the cis-cis, cis-
trans and trans-trans; the fourth possibility, trans-cis, is identical to the cis-
trans isomer. A general system of nomenclature has recently been developed for
olefins of the type abC=Ccd. Suppose that, according to the sequence rule
developed for enantiomers (see above) a precedes b and c precedes d. Then, if
a and c are cis to each other, the configuration is said to be seqcis (= sequence
cis) and the symbol Z (for German zusammen = together) is placed before the
name of the compound. Contrariwise, if a and c, the groups of higher precedence
on each side, are trans to each other, the configuration is seqtrans and the symb
is E (for German entgegen = across). It should be noted that in a few cases
- such as $\underset{H}{\overset{Cl}{>}}C=C\underset{Br}{\overset{Cl}{<}}$ - seqcis and seqtrans do not correspond to the classica
cis and trans: the cis-1,2-dichloro-1-bromoethylene shown must be designated
seqtrans (Cl precedes H on left, Br precedes Cl on right, Br and Cl are trans)
and its alternative name is (E)-1,2-dichloro-1-bromoethylene.

The determination of configuration of cis-trans isomers is very much simple
than that of enantiomers. For example, in the case of the butenedioic acids
(Fig. 12), the cis isomer (maleic acid) distinguishes itself by readily forming a

cyclic anhydride; hydration of the anhydride returns the acid. Now rings of 3-6 members can exist with cis double bonds only; it takes at least five atoms to span the trans positions of the C-C double bond even fleetingly and a six-atom bridge (as in trans-cyclooctene) is required to produce a stable structure.

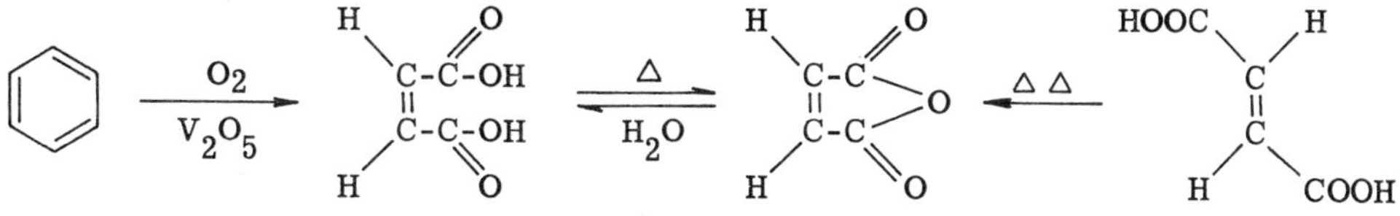

	Maleic acid	Maleic anhydride	Fumaric acid
	cis	cis	trans

The Butenedioic Acids

Fig. 12

Therefore maleic anhydride and thus also maleic acid have a cis double bond; the isomeric trans isomer, fumaric acid, does not form an anhydride easily and, under stringent conditions, is converted to maleic anhydride, presumably by prior isomerization about the double bond which is known to occur at elevated temperatures in many olefins, even though the barrier to rotation about double bonds (of the order of 40 kcal/mole) is very much higher than the barrier in saturated compounds, such as ethane or butane (see above). Further evidence for the cis configuration of maleic acid comes from the fact that the acid is obtained by oxidative degradation of benzene in which the double bonds are also necessarily cis. Having once established the configuration of a few molecules, such as maleic acid, one can then obtain that of others by correlative methods, using the principle of least structural change, in a manner similar to that used in the configurational assignment of enantiom ers. For example, the configuration of trans-crotonic acid, $CH_3CH=CHCOOH$ is determined by correlating it with trans-trichlorocrotonic acid $CCl_3CH=CHCOOH$ from which it is obtained by hydrogenation of the CCl_3-group to CH_3; the configuration of trans-trichlorocrotonic acid, in turn, follows from the fact that it is hydrolyzed to fumaric acid (Fig. 12 $CCl_3 \longrightarrow COOH$).

cis-trans Isomers usually differ substantially in physical properties. For example, the dipole moment of trans-1,2-dichloroethylene, HClC=CHCl is zero, since the C-Cl dipoles are opposed and cancel each other, but the dipole moment of the cis isomer is 1.89 D. (This, incidentally, provides an additional means for assigning configuration.) Other differences are in melting point (generally higher for the trans isomer), boiling point, refractive index and density (generally higher for the isomer of higher dipole moment), acidity, ultraviolet, infrared and NMR spectra, etc.

With the 2-butenes, $CH_3CH=CHCH_3$ and other carbon-substituted olefins, such as crotonic acid, $CH_3CH=CHCOOH$, the trans isomer is more stable than the cis. Probably this has reasons of steric origin, for, as may be seen in Fig. 12, the cis-substituents in a disubstituted olefin are close together and may crowd each other. The relative stability of the isomers may be established by comparing their heats of combustion (higher for cis-alkenes than for trans, because of the higher energy content of the former), heat given off during hydrogenation to (the same) paraffin (also higher for cis than for trans) or isomerization of the two configurational isomers, either thermally or catalytically (the more stable trans isomer predominates at equilibrium). When one looks at other types of olefins however--for example, 1,2-dihaloolefins, XHC=CHX (X = halogen) or 1-halopropenes, XHC=CHCH$_3$--it is no longer true that the trans isomer is necessarily the more stable; in fact, in the last two cases the cis isomer usually predominates at equilibrium.

c) Allenes. Cumulenes. Contemplation of a molecular model of an allene (1,2-diene) of the type HXC=C=CHX shows that the planes of the substituents at the two ends are at right angles to each other (Fig. 13). It follows that allenes of this type are devoid of a plane of symmetry; i.e. they are dissymmetric or chiral and may, in principle, be resolved into enantiomers. While this was recognized by van't Hoff in 1874 it was only 61 years later (in 1935) that Maitland and Mills in England and Kohler, Walker and Tishler in the United States obtained the allenes $C_6H_5(\propto$-$C_{10}H_7)C=C=CC_6H_5(\propto$-$C_{10}H_7)$ $(\propto$-$C_{10}H_7 =\propto$-naphthyl) and $C_6H_5(\propto$-$C_{10}H_7)C=C=CC_6H_5(COOCH_2COOH)$ in optically active form.

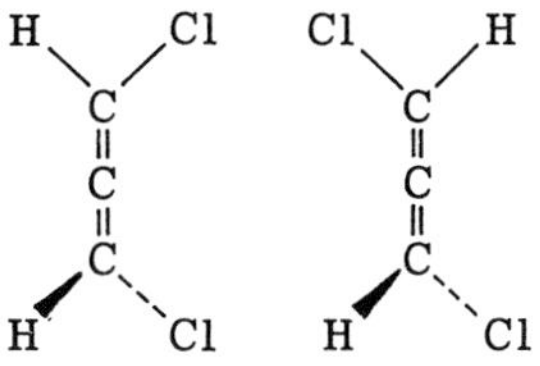

1,3-Dichloroallene-
enantiomers

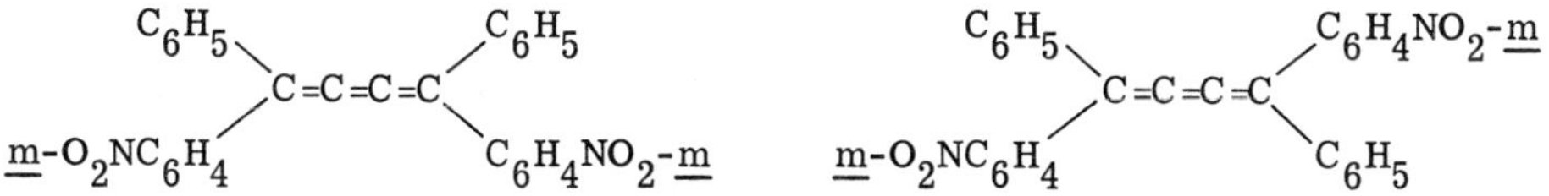

cis-trans Isomers of 1,4-diphenyl-1,4-
di-(m-nitrophenyl)butatriene

Allenes and cumulenes

Fig. 13

Allenes do not contain a chiral center but provide examples of a different source of dissymmetry, the so-called "chiral axis", in this case the C=C=C axis. Other cases of axial chirality will be discussed below.

When there is an additional double bond, as in a "cumulene" abC=C=C=Cab, the groups a and b will again be in plane and the molecule exhibits cis-trans isomerism rather than enantiomerism (Fig. 13). In general, an odd number of cumulated double bonds will give rise to cis-trans isomerism and an even number to enantiomerism.

d) Acetylenes. A model of an acetylene, R-C≡C-R' shows the molecule to be linear. (Doubly-bonded carbon atoms may be represented as two tetrahedra sharing a common edge and triply-bonded carbons as two tetrahedra joined at a face. This formal representation--or the similar representation in a ball-and-stick molecular model set in which the multiple bonds are represented by bent springs--may actually bear a fairly close resemblance to the true quantum-mechanical nature of the multiple bonds.) Because of the linearity of acetylenes, they show no stereochemical features in their structure unless it be in the attached groups R or R'.

e) Cycloalkanes. Strain. When the two ends of a hydrocarbon chain are joined in a ring, a cyclic hydrocarbon or cycloalkane results. There are three important differences between cycloalkanes and open-chain alkanes. First of all, in the small cycloalkanes there is a considerable deviation of the C-C-C bond angle from the normal tetrahedral value of $109^{o}28'$. For example, in cyclobutan C_4H_8, a square molecule, the angle is 90° and in cyclopropane, C_3H_6, a triang shaped molecule, the angle is 60°. This deviation from the normal bond angle causes what is called "angle strain"--it takes energy to deform the normal tetr hedral bond angle in such a way that the small ring may be closed. This strai (also called "Baeyer strain" after its discoverer, Nobel Laureate Adolf von Baeyer) may be measured quantitatively as the difference in heat of combustion of the ring in question and of a methylene chain $(CH_2)_n$ of the same number of carbons. The latter heat of combustion is arbitrarily taken as n x 157.4 kcal/mol, 157.4 kcal/mol representing the average increment in heat of combustion in going from a hydrocarbon C_nH_{2n+2} to its next higher homologu $C_{n+1}H_{2n+4}$. For cyclopropane, the heat of combustion is 499.8 kcal/mol and the strain therefore 499.8 - 3 x 157.4 or 27.6 kcal/mol (corresponding to one-third of this amount, or 9.2 kcal/mol per methylene group). The corresponding value for cyclobutane are: heat of combustion, 655.80 kcal/mol, total strain, 26.2 kcal/mol, strain per CH_2-group, 6.55 kcal/mol. While the total strain for cyclobutane is similar to that for cyclopropane, the strain per methylene group is appreciably less, reflecting the much lesser deformation of the valence angle in cyclobutane (from $109^{o}28'$ to 90°) as compared to cyclopropane ($109^{o}28'$ to 60°).

A second difference between open-chain alkanes and cycloalkanes is that the chain segments in alkanes are always staggered and usually in the anti conformation, whereas cycloalkanes, of necessity, have some segments in the gauche conformation or even in the cis (eclipsed) conformation in order that the ring may be closed. Cyclopentane is an example: the measured strain in this molecule is 793.5 - 5 x 157.4 = 6.5 kcal/mol or 1.3 kcal per methylene group. Although this is considerably less than the strain in cyclopropane or even cyclo-butane, it is at first glance surprising that cyclopentane should have any strain at all since the angle in a regular pentagon--108°--is so close to the tetrahedral angle that angle strain is negligible. (Angle strain is approximately equal to $0.017 \, \Delta\theta^2$ kcal/mol where $\Delta\theta$ is the deviation (in degrees) from the normal valency angle. Thus the strain is only 0.07 kcal per bond angle for a deviation of 2°, 1.7 kcal for 10°, 6.8 kcal for 20°, the approximate deformation in cyclo-butane.) However, in the planar, pentagonal form of cyclopentane, model considerations indicate that all the C-C and C-H bonds are eclipsed and as a result a great deal of eclipsing or "torsional strain" (sometimes also called Pitzer strain, after its discoverer, Kenneth Pitzer) is incurred. If cyclopentane were a planar pentagonal molecule, this strain would amount to 10-15 kcal/mol. In fact, however, the cyclopentane molecule is somewhat puckered; in this way it avoids much of the eclipsing strain and even though some angle strain is created, the total strain (angle strain plus torsional strain) is reduced to 6.5 kcal/mol.

In cycloheptane, the situation is similar to that in cyclopentane and the strain is 1108.0 - 7 x 157.4 = 6.2 kcal/mol or ca. 0.9 kcal per CH_2 group. In cyclohexane, however, there is no strain; the heat of combustion, 944.5 kcal/mol, is almost exactly that calculated (6 x 157.4). This is because of the perfectly staggered chair shape of the cyclohexane molecule; it may be seen in models that in this conformation all bond angles are the normal tetrahedral angles and all hydrogen atoms are perfectly staggered. (Actually, the situation is almost but not quite as described; since the C-C-C bond angle is somewhat larger--112°-- and the H-C-H angle somewhat smaller--108°--than tetrahedral, the cyclohexane ring is, in fact, a somewhat flattened chair.)

In cyclooctane, a new factor appears. Model considerations indicate that some of the hydrogen atoms point toward the center of the ring and begin to interfere sterically. There is thus yet another, new, source of strain, the so-called "van der Waals strain" or "non-bonded interaction". This type of interaction increases the strain per CH_2 group in cyclooctane to 1.2 kcal (total strain 9.7 kcal/mol, based on a heat of combustion of 1268.9 kcal/mol). Strains of this magnitude (1.4 kcal/CH_2 for cyclononane, 1.2 kcal for cyclodecane, 1.0 kcal for cycloundecane) persist from eight- up to eleven-membered rings; after that the strain drops sharply, to 0.3 kcal/CH_2 for the twelve-membered ring and finally to near zero for rings of fourteen members and larger where once again one has normal bond angles, staggered bonds and no interference of groups across the ring.

In summary, then, rings fall into four classes. High angle strain exists in the so-called "small rings" (3- and 4-membered). "Common rings" (5-, 6- and 7-membered) have low angle strain and varying but small amounts of eclipsing strain, the total strain being small. Eclipsing strain is minimized in 5- (and even 4-) membered rings through puckering. (All rings larger than five-membered are also non-planar.) "Medium rings" (8- to 11-membered) are subject to non-bonded interactions across the ring (giving rise to "transannular strain" from Latin trans = across and annulus = ring); since this type of strain can be quite severe, the rings become appreciably deformed by ring-angle expansion (for example, the mean C-C-C angle in the ten-membered ring is 116.5^{o}) but some non-bonded strain persists in addition to angle strain and torsional strain and the total strain is quite appreciable. "Large rings"--12-membered and larger--no longer show this type of strain and are comparable in heat of combustion to linear chain segments of the same number of carbon and hydrogen atoms.

When one considers the ease of formation of ring compounds (as distinct from their stability), another factor, in addition to strain, must be taken into account, namely the probability of making the ends of a chain approach so as to form a ring. Clearly this probability decreases as the number of chain-members increases, i.e. it is highest for a three-membered ring (there is only one possible staggered conformation for a C_3-chain) and decreases from there on. The interplay of the probability factor and the strain factor produces some curious results, thus, for example, a three-membered ring is more easily formed than a four-membered one (because of the probability factor) but is also less stable and more easily broken (because of the strain factor). The same is true for five- vs. six-membered rings, the former being generally easier to form but the latter being more stable. From seven to eleven members, the strain and probability factors conspire to make ring formation difficult and rings of 9, 10 and 11 members were, in fact, virtually unknown until 1947 when Hansley, Prelog and Stoll found that a special type of synthesis, the so-called acyloin reaction, was peculiarly suitable for the preparation of rings of this size. The synthesis of the nearly strain-free twelve- and larger member rings is actually somewhat easier, although, in view of the low probability of bringing the ends of a chain of twelve atoms or longer together, it is by no means straightforward and Leopold Ruzicka in Zurich, Switzerland, won the Nobel Prize in 1939 for first devising methods (in the 1920's) for synthesizing large ring compounds, including the naturally occurring perfumes muskone (from musk deer) and civetone (from the civet cat). (The name of another Nobel laureate, Karl Ziegler, is also associated with the synthesis of large ring compounds.)

So far we have considered only simple cycloalkanes $(CH_2)_n$. A few considerations regarding substituents or ring elements different from CH_2-groups are in order. In general, it is found that geminal substitution (i.e. presence of two substituents on the same atom--from Latin geminus = twin) aids ring formation; thus $(CH_3)_2C(CH_2Cl)_2$ (A) will cyclize to

$(CH_3)_2C\begin{smallmatrix}\diagup CH_2 \\ | \\ \diagdown CH_2\end{smallmatrix}$ more readily than $CH_2(CH_2Cl)_2$ (B) cyclizes to cyclopropane.

This effect, the so-called "Thorpe-Ingold effect" (named after its discoverers), depends on bond angles: In (A) above, for reasons of symmetry, the CH_2Cl-groups will be at an angle of nearly $109°28'$ to each other and can therefore be brought together more readily than in (B) where the $ClCH_2-C-CH_2Cl$ angle is closer to $112°$. (A different explanation has been advanced for larger rings.) When there are "hetero-atoms" (such as oxygen, sulfur, nitrogen; Greek heteros = different, meaning atoms different from carbon) in the ring to be formed, ring formation is generally easier than for carbocyclic rings (i.e. rings containing only carbon), especially in the medium-ring region. This is because the major non-bonded repulsion in medium rings is between hydrogens; thus when the hydrogens are absent (for example, when a $-CH_2-$ is replaced by $-O-$) the strain is decreased and ease of ring formation is increased.

A study has been made of rings containing double and triple bonds. A double bond may be incorporated even in the three-membered ring: cyclopropene is known despite its great strain. The geometry of substitution of the double bond must be cis in small rings; cyclooctene is the smallest ring which can accomodate a trans-substituted double bond (although trans-cycloheptene and trans-cyclohepten-2-one have been obtained fleetingly, as reaction intermediates). trans-Cyclooctene is appreciably strained (by 9.3 kcal/mol more than the cis isomer) and it is so stiff a molecule that it exists in enantiomeric forms which are not readily interconverted (except on heating); thus the molecule may be resolved, in contrast to simple olefins (which are planar) or cyclohexene (which is chiral but suffers interconversion of its enantiomers at a rate so high that resolution is impossible).

The smallest ring in which a triple bond may be accommodated is also eight-membered (cyclooctyne).

f) Alkylidenecycloalkanes. Spiranes. In allenes (see above) we have seen an example of dissymmetry involving a chiral axis. Related molecules are alkylidenecycloalkanes (first resolved by Perkin, Pope and Wallach in 1908) and spiranes (from Latin spira = pretzel) first resolved by Mills and Nodder (1920) shown in Fig. 14. The absolute configuration of representative alkylidenecyclo-alkanes and spiranes has been established by correlation with compounds having chiral centers, as has that of several substituted allenes, discussed earlier.

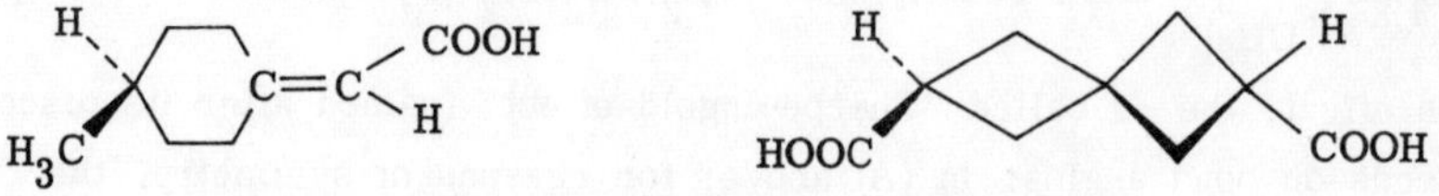

Resolvable alkylidenecycloalkane and spirane

Fig. 14

The system of configurational nomenclature for allenes, alkylidenecyclohexanes and biphenyls (see below)--all compounds with chiral axes -- cannot be discussed here but is described in the earlier-cited article by Cahn, Ingold and Prelog, as is that for spiranes.

g) Biphenyls. The molecule of biphenyl (Fig. 15) exists in a variety of conformations generated by rotation about the bond between the aromatic rings.

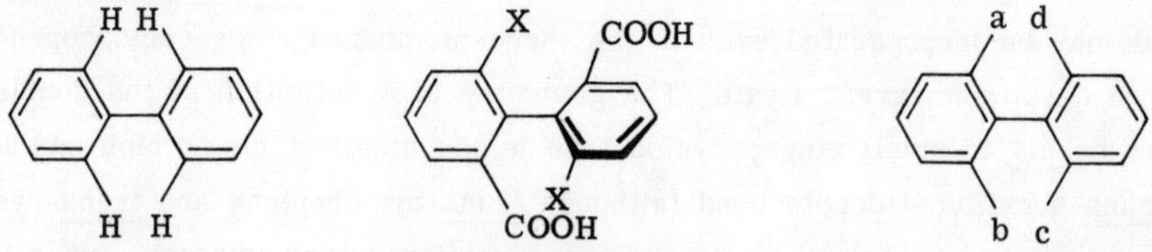

Biphenyl

X = H   : Diphenic acid
X = NO$_2$: 6,6'-Dinitrodiphenic acid

Generalized formula

Biphenyls

Fig. 15

A priori, it is somewhat difficult to predict the position of the conformational minimum in this case, for whereas the planar form of the molecule provides for maximum overlap of the pi-electrons of the aromatic rings (and therefore resonance stabilization), it suffers from steric interaction between the ortho hydrogens (shown in Fig. 15). This interaction would be minimized in a conformation in which the rings are at right angles, but in this conformation resonance between the aromatic rings would be totally inhibited. Apparently a compromise is struck, for electron diffraction data of biphenyl vapor suggest that the rings are at a 40° angle; the ultraviolet spectrum of biphenyl indicates that a considerable amount of resonance interaction between the two rings persists at this angle. There is every indication that biphenyl rotates about the Ar-Ar axis relatively easily, passing, however, through an energy maximum at or near the coplanar conformation and through another maximum at or near the conformation where the rings are at right angles. When the rings are substituted in two of the ortho positions (as in diphenic acid, Fig. 15), the picture changes but in a minor way. The angle between the rings at the energy minimum is now increased (to over 60°), the perpendicular energy barrier becomes lower and the coplanar barrier higher. The ultraviolet spectrum now shows greatly diminished resonance interaction between the two aryl rings. It may also be noticed that the two possible perpendicular conformations are no longer superimposable but are enantiomeric. The situation in biphenic acid is thus best described as that of two rapidly interconverting enantiomers. When the size of the substituents in the ortho positions is increased still more, as in 6,6'-dinitrodiphenic acid (Fig. 15), the energy of the conformation in which the rings are coplanar now becomes quite high (because of non-bonded interactions of the ortho substituents in both possible coplanar arrangements) and therefore a substantial barrier (of the order of 20 kcal/mol or more) is set up to interconversion of the two enantiomers, the stable (enantiomeric) conformations now corresponding to a nearly perpendicular arrangement of the rings. 6,6'-Dinitrodiphenic acid is therefore a resolvable molecule, as was discovered by Christie and Kenner in England in 1922. Whereas we normally speak of conformational isomerism in biphenic acid, the stable enantiomers of its 6,6'-dinitro derivative may be said to be of opposite configuration (in analogy with (+) and (-)-lactic acid), although it is equally correct to say that they differ in conformation. The term "atropisomerism" (from Greek a = not and tropos = turning) has also been applied to this type of stereoisomerism.

Atropisomerism depends on the existence of a chiral axis passing along the pivotal bond and through the 4 and 4' atoms in the two phenyl rings. One condition for the existence of such a chiral axis is the absence of a plane of symmetry in the perpendicular conformation shown in Fig. 15 (general formula. This condition will pertain if a ≠ b and c ≠ d, but even if a = b (or c = d) may be achieved by placing a group in the meta position of that ring in which the ortho substituents are equal. A second condition is that there is a high enough barrier at the planar conformation to prevent the enantiomers from interconverting rapidly by passing through this conformation. Clearly this is no an all-or-nothing proposition: if the barrier is small (less than 16 kcal/mol), resolution will be impossible at room temperature; if the barrier is high (over 25 kcal/mol), it will be possible and the enantiomers will be stable; if it is intermediate, resolution may be possible but the resolved enantiomers will racemize on standing or heating by gradually passing through the planar conformation. In general, resolution is not possible unless at least one of the ortho-substituents in each ring (say a and c) is larger than hydrogen; even the a relatively large substituent is required (for example a = c = Br, b = d = H) and racemization will occur rather easily. With three ortho substituents (a, b, c ≠ H, d = H) resolution is generally possible unless one of the substituents is fluorine (small) but the compounds are racemized on heating to temperatures i the vicinity of 120° for periods ranging from a few minutes to a few days. Te ortho-substituted biphenyls (a, b, c, d ≠ H) are generally resolvable and often qu hard to racemize, except when all the o-substituents are fluorine or methoxyl (which are too small to prevent rotation of the rings through the all-planar conformation).

A considerable amount of work has been done in the area of biphenyl dis symmetry. The absolute configuration of biphenyls has been correlated with th of compounds with chiral centers (ultimately, lactic acid), a system of configurational nomenclature has been developed for the biphenyls, good theoretical predictions of racemization rates in substituted biphenyls have been developed and the stereoisomerism of biphenyls with bridges across the ortho and ortho' positions has been studied.

Related to biphenyl atropisomerism is the atropisomerism of other sterically crowded compounds, such as the styrene shown in Fig. 16.

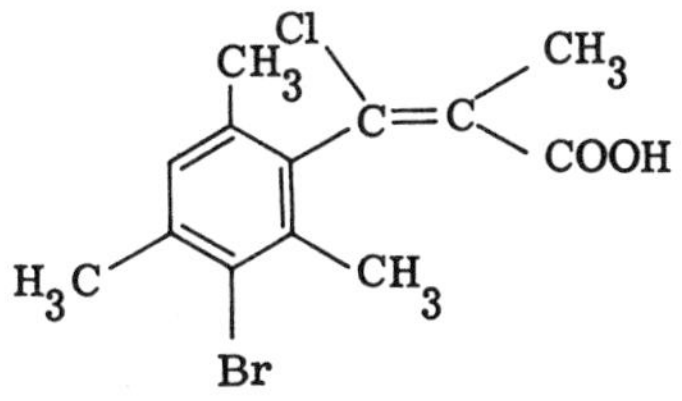

Hindered rotation in styrene

Fig. 16

h) Configuration and Conformation. Configurational differences have been described earlier as giving rise to either enantiomers or diastereoisomers. Conformational differences may also give rise to enantiomers and diastereoisomers; thus the two possible gauche forms of 1,2-dibromoethane (Fig. 11) are enantiomers but the anti form is a diastereoisomer of either gauche form. These considerations do not therefore define a distinction between configuration and conformation. Conformation has earlier been described as being based on rotation about single bonds; rotation about double bonds (as in cis and trans disubstituted ethylenes, Fig. 12) is considered to give rise to configurational, not conformational isomers because the barrier is very much higher than the barriers to rotation about single bonds. Unfortunately, as the case of the biphenyls demonstrates, there is not a large area of exclusion between the situations in singly-bonded and doubly-bonded compounds but rather a continuum: from ethane (barrier ca. 3 kcal/mol) to cyclohexane (barrier ca. 11 kcal/mol, see below) to biphenic acid (barrier about 15 kcal/mol) to 2,2'-diiodobiphenyl (21 kcal/mol) to 6,6'-dinitrobiphenic acid (ca. 30 kcal/mol) to 2-butene (ca. 40 kcal/mol). There is therefore no definite point where one can say that conformational isomerism ends and configurational isomerism begins, but rather there is an area of transition comprising not only the variously substituted biphenyls but also certain other compounds where there is restricted rotation around single or partial double bonds, as for example in thioamides

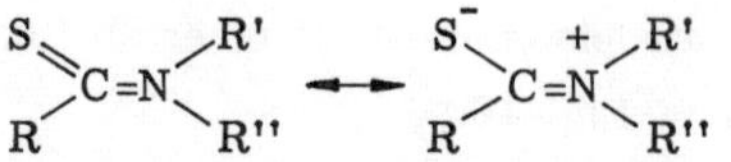

$$S=C-N\begin{smallmatrix}R'\\R''\end{smallmatrix} \longleftrightarrow S^{-}-C=N^{+}\begin{smallmatrix}R'\\R''\end{smallmatrix}$$ which have been isolated in <u>cis</u> and <u>trans</u>

forms (Walter, 1963, 1968).

Since rate of rotation about bonds is a function of temperature as well as system and since the chances of distinguishing conformational isomers depends not only on their lifetime but also on the methods used for their detection, the situation is even more complex than that. Take, for example, chlorocyclohexane As seen in Fig. 17, this molecule exists in two conformations, the Cl-equatorial

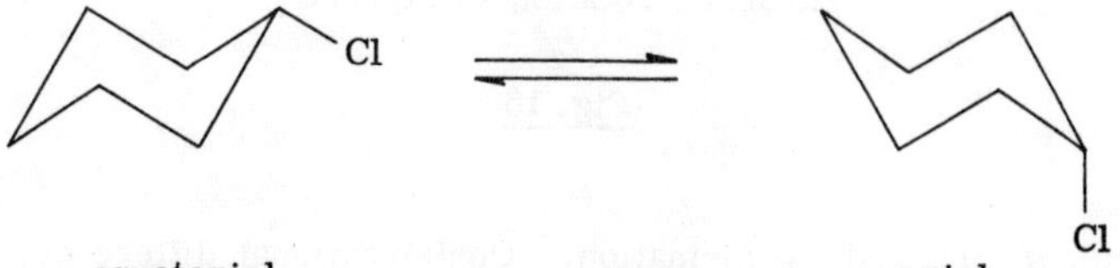

equatorial   axial

Chlorocyclohexane

<u>Fig. 17</u>

and the Cl-axial. The equatorial conformation is more stable by about 0.4 kcal/m (i.e. 2 molecules out of three at room temperature will have equatorial chlorine and the barrier between the two conformations is <u>ca.</u> 11 kcal/mol, meaning that the rate of interconversion of the chair forms at room temperature is of the order of 100,000 times per second. It follows that not only can the two conformational isomers not be separated, but they cannot either be seen individually in the nuclear magnetic resonance spectrum but rather an average of the two conformations is seen. However, both conformations are seen in the infrared spectrum which displays distinct frequency bands for the equatorial and axial C-Cl stretching vibrations. When the temperature is lowered to about -75°C, the situation changes (Jensen, 1960). The rate of interchange of the two isomers is reduced to about 10 interconversions per second which is sufficiently slow so that the NMR spectra of the two isomers can now be seen distinctly. Finally, if the temperature is lowered to -150°C, the rate of interconversion of the two isomers is so slow that they become physically isolable. The equatorial isomer, in fact, crystallizes out (Jensen, 1966) and, if redissolved at -150°, displays its own NMR spectrum, whereas the axial isomer, of distinct NMR spectrum, remains in the mother liquor. Thus, if isolation is taken as the distinguishing criterion, it may be said that at -150° chlorocyclohexane displays configurational isomerism but at -75° and +25° it displays conformational isomerism; NMR spectroscopy is capable of distinguishing the conformational isomers at -75° but not at +25°.

Clearly there is a barrier area, somewhere between 10 and 20 kcal/mol where the difference between configurational and conformational isomerism is not sharply defined; in this area each investigator must make his own choice of terminology, state what it is and adhere to it consistently.

i) Planar Chirality. Ansa Compounds. Cyclophanes. We have seen that enantiomerism can originate from a dissymmetrically substituted atom (chiral center; however, a chiral center need not always coincide with an atom--note,

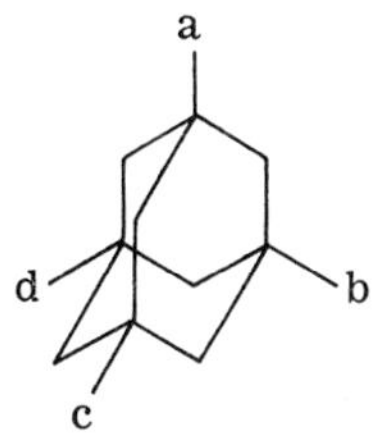

| Chiral adamantane derivative | Cyclohexene | trans-Cyclooctene |

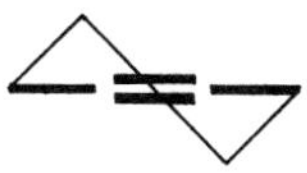

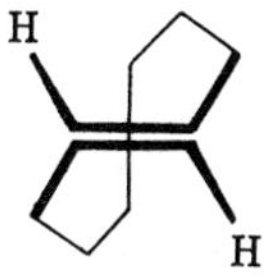

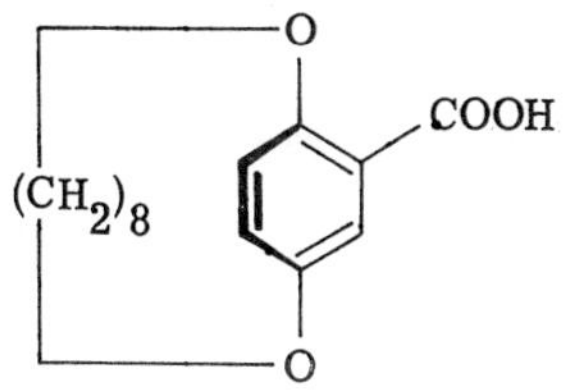

Ansa compound 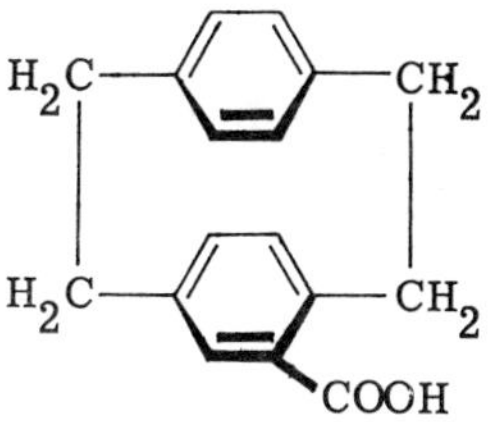   Paracyclophane

Miscellaneous chiral molecules

Fig. 18

for example, the case of the tetrasubstituted adamantane shown in Fig. 18) or from dissymmetric substitution about a "chiral axis" as in biphenyls, allenes, etc. There is a third case where neither a point nor an axis can be defined as the focus of chirality, but rather a chiral plane can be defined, the dissymmetry being due to different atoms or groups being located on the two sides of the plane. Cyclohexene (Fig. 18) is a simple case, but here the two enantiomers are readily interconverted (barrier ca. 6 kcal/mol) and so are conformational, not configurational isomers. Configurational enantiomers with a chiral plane are, however, found (Cope, 1963) in the already mentioned trans-cyclooctene (Fig. 18 (the chiral plane here is the plane containing the double bond). Other cases are the "ansa compounds" (from Latin ansa = handle) shown in Fig. 18 (Lüttringhaus, 1940) and the "paracyclophanes" (Fig. 18, Cram and Allinger, 1955); in both these cases planar chirality is maintained because the aromatic ring or rings are sterically prevented from passing through the plane of the larger ring.

j) Helicity. 4,5-Disubstituted Phenanthrenes. There are some molecules whose enantiomerism is best defined in terms of helicity. A helix, of course, is an inherently dissymmetric entity, being either right-handed (twisting clockwise away from the observer) or left-handed (twisting counterclockwise away from the observer). Right- and left-handed helices of the same shape are mirror images (enantiomers); corresponding to the symbols (R) and (S) used for chirality are symbols (P) (for plus) and (M) (for minus) used for right-handed and left-handed helicity.

An example of helicity is found in the 4,5-disubstituted phenanthrene shown in Fig. 19. This molecule was resolved by Newman (1947); due to severe non-bonded interaction of the substituents attached at positions 4 and 5, the aromatic system is non-coplanar, i.e. twisted, imparting a helical sense to the molecule. A much more optically stable system is hexahelicene (Fig. 19: Newman, 1956); it is clear that this molecule would present a most severe steric interference of the non-bonded edges of the top carbon rings if it were planar. In fact, the molecule possesses a strong helical twist and, once resolved, shows the remarkable specific rotation of $3,700^{\circ}$. Higher homologs (hepta-, octa-, nona-helicene) are also known (Martin, 1968). A somewhat different helical molecule of much lower optical stability shown in Fig. 19 is tri-o-thymotide which upon crystallization forms enantiomorphous crystals of opposite chirality; when dissolv

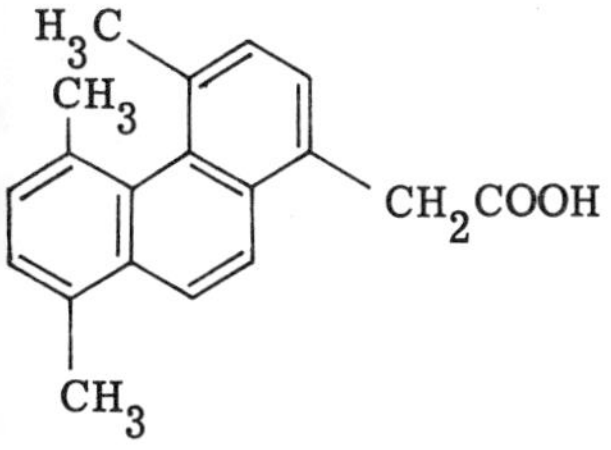

4,5,8-Trimethylphen-   Hexahelicene 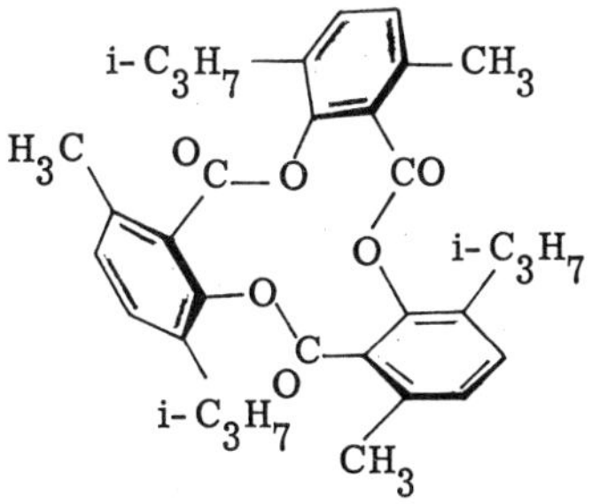   Tri-o-thymotide
anthrene-1-acetic acid

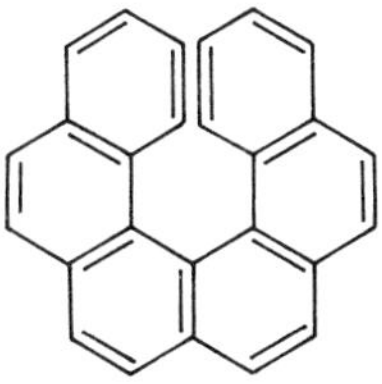

Examples of helical dissymmetry

Fig. 19

the material racemizes rapidly by interconversion of the two helical forms by twisting of the aromatic rings. (The material may also function as a resolving agent, for when crystallized from certain chiral solvents, such as 2-bromo-butane, $CH_3CHBrCH_2CH_3$, it forms diastereoisomeric crystals of an "inclusion compound" containing one enantiomer of the bromobutane contained within the crystal lattice of one enantiomer of the thymotide. Separate crystals containing the two opposite enantiomers are, of course, also formed.)

Some very important naturally occurring polymers or macromolecules, such as proteins, polypeptides and nucleic acids form molecular strands which are partly or completely helical.

k) Catenanes. A particularly intriguing class of molecules, from the point of view of three-dimensional chemistry, is presented by the catenanes, shown in

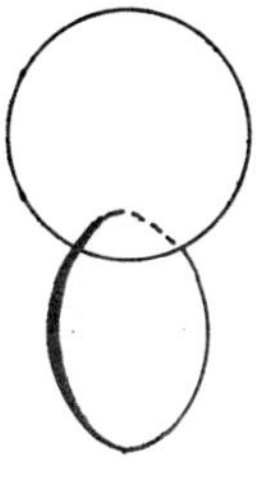

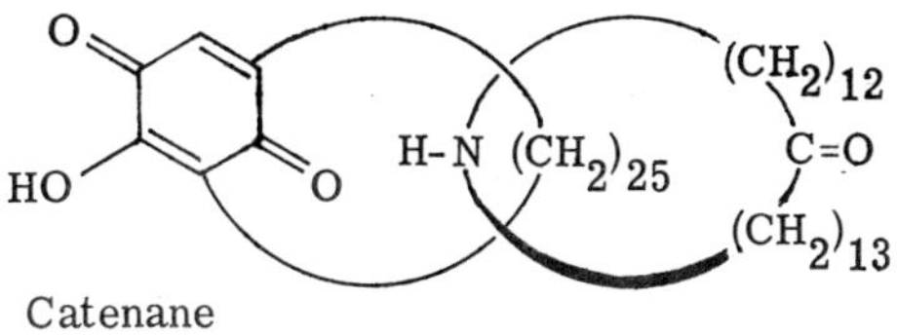

Catenane

Fig. 20

Fig. 20, in which two rings are topologically interlocked. A well-documented example of a catena compound (Fig. 20) has been prepared by rational synthesis (Lüttringhaus and Schill, 1964).

V. Stereochemistry of Other Tetrahedral Elements

a) Quadriligant Compounds. A number of elements other than carbon form tetrahedral valencies and appropriate compounds of structure Xabcd have been obtained in optically active form. Quaternary ammonium salts of type Nabcd$^+$ X$^-$ were resolved by Pope and Peachey (1899) and constitute the first known example of chirality not due to carbon. Amine oxides of type abcN$\rightarrow$O have been resolved by Meisenheimer (1908). Doubly bonded nitrogen compounds, such as oximes RR'C=NOH exist as cis-trans diastereoisomers (Fig. 21; the terms syn and anti are generally used for oximes in lieu of cis and trans) similarly to olefins RR'C=Cab; it is to be noted that in the oxime, a pair of electrons occupies one of the four positions occupied by an atom (a or b) in the case of the analogous olefin. However, stable enantiomers of tri-substituted acyclic amines of type abcN: (where again a pair of electrons occupies the fourth coordination site) are not known. It has been shown that, although amines abcN: are pyramidal (i.e. tetrahedral with one corner of the tetrahedron occupied by the free electron pair) a very rapid inversion or involution of the pyramid may occur converting the molecule to its enantiomer (Fig. 21) in other words, racemization is so fast that resolution cannot ordinarily be achieved. (The case is similar to that of diphenic acid (Fig. 15) but the reasons for rapid racemization are different: rapid rotation in the case of diphenic acid, rapid inversion or involution in the case of tertiary amines.) Exceptions have recently been found in N-chlorethyleneimines (Brois, 1968; Eschenmoser, 1968); for example, the N-chloropropyleneimine shown in Fig. 21 has been obtained in cis and trans forms (which are relatively easily interconverted at room temperature).

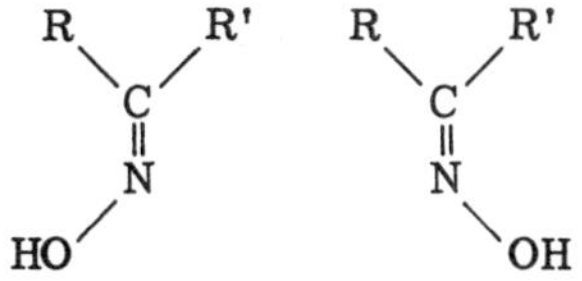

syn-R   anti-R

Oximes   Tertiary amines (not resolved)

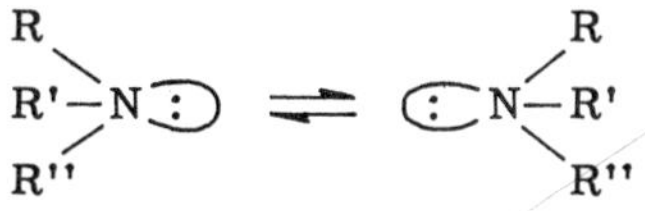

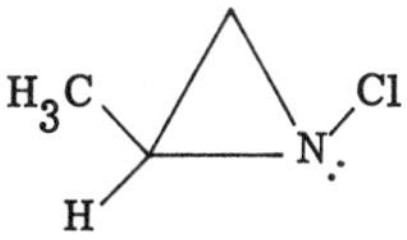

cis 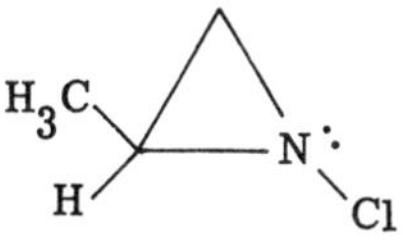   trans

N-Chloropropylene imines

Stereochemistry of nitrogen

Fig. 21

Of the other elements of the first row of the periodic system boron and beryllium have been found to give rise to resolvable compounds in which the central atom is "quadriligant" (i. e. has four atoms or groups tied to it--Latin ligare = to tie). Quadriligant compounds of lithium, oxygen, fluorine and neon, the remaining elements of the first row, are not known.

Of the elements of the second row, silicon, phosphorus and sulfur are known to give rise to optically active compounds. Optically active silicon compounds of the type Siabcd were discovered by Kipping (1908; for a detailed account of recent work, see L. H. Sommer, "Stereochemistry, Mechanism and Silicon", McGraw-Hill Book Co., Inc., New York, 1964) and a number of examples have been studied. Phosphine oxides, $Pabc \rightarrow O$, were resolved by Meisenheimer (1911) and a number of active phosphonium salts, $Pabcd^+ X^-$, have also been prepared. Further down the periodic table, optically active Germanium compounds, Geabcd, tin compounds, Snabcd, and arsonium salts, $Asabcd^+ X^-$, are known.

b) Triligant Compounds. It has already been mentioned that amines with three ligands (in contrast to quaternary ammonium salts with four ligands) cannot be resolved because of rapid spontaneous inversion (Fig. 21). However, as one progresses from nitrogen to the second row of the periodic system, such vibrational inversion becomes very much less rapid and optically stable derivatives are known of triligant phosphorus, :Pabc, arsenic, :Asabc, and antimony, :Sbabc. Optically stable derivatives of triligant sulfur have also been studied extensively; foremost among them sulfonium salts, :Sabc$^+$ X$^-$, and sulfoxides abS̈ → O. Analogous salts of selenium, :Seabc$^+$ X$^-$, and tellurium, :Teabc$^+$ X$^-$, have similarly been obtained in optically active form.

c) Triligant Derivatives of Carbon. Nearly all stable derivatives of carbon are either quadriligant or multiply bonded. There are, however, three important triligant reaction intermediates derived from carbon compounds: the carbonium ion, Cabd$^+$, the carbon radical, ·Cabd and the carbanion, :Cabd$^-$. While these species are fleeting, their stereochemistry has been of interest since it often determines the overall stereochemistry of the reactions in which they are intermediates (see dynamic stereochemistry below). The three species differ, apart from charge, by the fact that the carbonium ion has six valence electrons around the central carbon, the carbon radical seven and the carbanion eight.

Carbonium ions can be shown by various lines of evidence to be planar. Among these lines of evidence are theoretical calculations (the most stable hybridization of a carbonium ion should be sp^2 with a vacant p orbital, and the sp^2 array of valence electrons is planar), comparison with boron compounds (Babc compounds have been shown by physical measurements to be planar; Cabc is "isoelectronic" with Babc, meaning that the electrons surrounding the nuclei occupy the same shells in the two species) and the fact that reactions known or believed to proceed via carbonium ions proceed more rapidly when the ion is so located in the remainder of the molecule that it finds it very easy to become planar, but go only very slowly when the ion is prevented by its location from becoming planar.

Carbon radicals, ·Cabc, have been held in the gaseous state or in solution for times long enough to carry out physical observations (spectroscopic or electron paramagnetic resonance measurements); such measurements indicate that the radicals are planar or nearly planar.

Carbanions, :Cabc$^-$, are isoelectronic with tertiary amines and, like tertiary amines, are pyramidal but lose their chirality rapidly through the type of molecular vibration shown in Fig. 21.

VI. Stereochemistry of the Remaining Elements

The stereochemistry of metal complexes (or coordination compounds) has been a subject of continued interest since the discovery of $CoCl_3 \cdot 6NH_3$ in 1789. This was followed by the preparation of many such compounds and by numerous attempts to explain the nature of these substances. The real break-through in our understanding of these materials was provided by Werner in 1893 (Z. anorg. Chem. 3, 267 (1893)). It is now recognized that most of the chemistry of metals deals with their complexes. These substances have many applications such as the essential role they play in certain biological reactions and in industrial processes such as the Ziegler-Natta Catalysis for the preparation of polyethylene (F. Basolo and R. G. Pearson, "Mechanisms of Inorganic Reactions", John Wiley and Sons, New York, 1967).

Metal complexes are rich in their stereochemistry. Whereas saturated carbon compounds are four-coordinated and have a tetrahedral structure, metal complexes are known with coordination numbers ranging from two through twelve, and these exhibit a variety of different structures. A brief account of four- and six-coordinated complexes is given and a few examples of compounds with other coordination numbers are listed.

a) Four-Coordinated Metal Complexes. Four-coordinated metal complexes have either tetrahedral or square planar structures. Tetrahedral complexes display, in theory, the same type of stereochemistry as that described above for tetrahedral carbon compounds. However, in practice it has not yet been possible to resolve the dextro and levo isomers of compounds of the type [Mabcd], where M is a metal and a,b,c, and d are different unidentate ligands. This is because such compounds are substitution labile (the groups a,b,c and d are readily replaced by solvent or other ligands) which permits rapid racemization (dextro $\rightleftharpoons$ levo interconversion).

If two of the individual ligands are linked forming a⌢b, then these form stable chelate rings with the metal. Metal chelates of this type are optically active (Fig. 22) and several such compounds have been resolved. This has been reported for compounds of Be(II), B(III), Zn(II) and Cu(II).

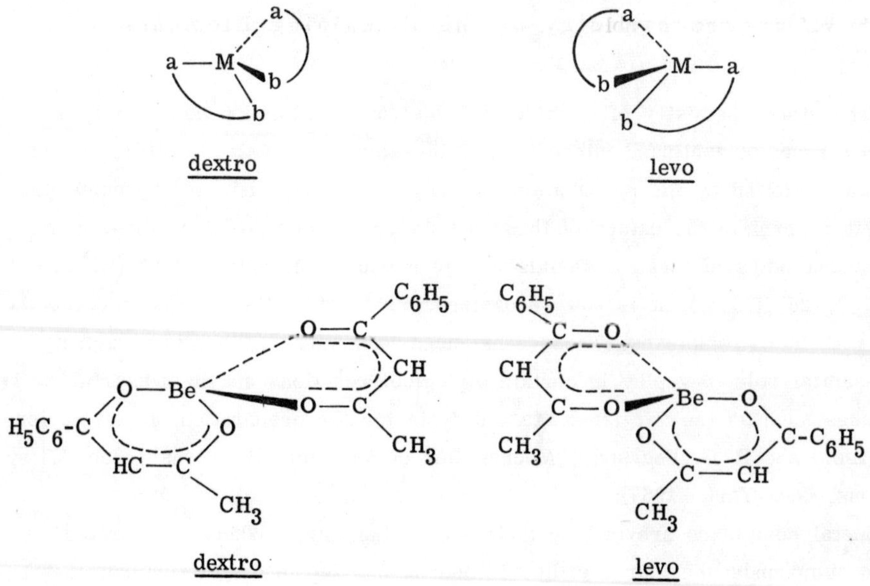

dextro   levo

dextro   levo

Enantiomers of tetrahedral metal complexes

Fig. 22

Except under rare and special circumstances, square planar complexes do not give rise to enantiomers, but they do form diastereoisomers (geometrical isomers) as, for example, in the case of compounds of the type [Mabcd] (Fig. 23)

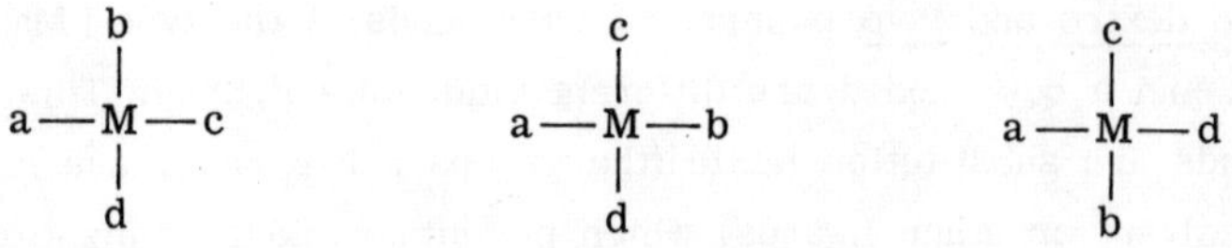

Possible geometrical isomers for a square planar [Mabcd] compound,   e.g.

[Pt(NH$_2$CH$_3$)NH$_3$ClBr]

Fig. 23

Only three forms are possible, one in which the a-M-c angle is 180° (trans positions), a second where this is the case for a-M-b and a third such with a-M-d. Platinum(II) forms compounds of this type, but much more common are $[Ma_2b_2]$ compounds. These can form only two isomers, the cis (like groups adjacent) and trans (like groups opposite) (Fig. 24). Many examples of such

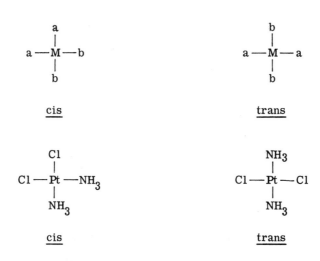

Geometrical isomers of $[Ma_2b_2]$ compounds

Fig. 24

isomers are known for Pt(II), and fewer are also known for compounds of Pd(II), Ni(II), Au(III), Rh(I) and Ir(I).

Square planar metal chelates $[M(a\frown b)_2]$ likewise do not give rise to enantiomers (Fig. 22) but form instead geometrical isomers (Fig. 25).

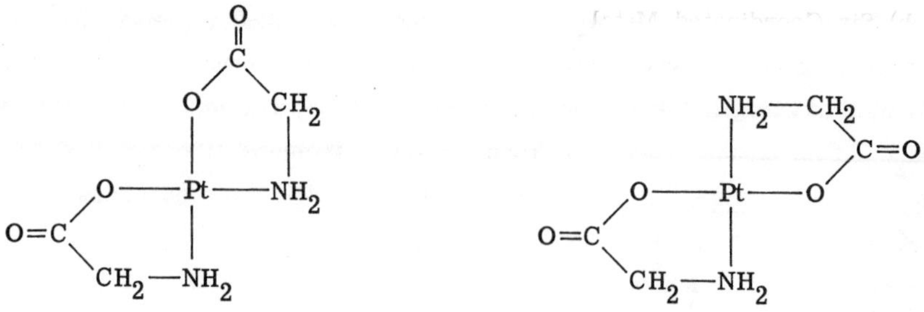

cis

trans

Geometrical isomers of metal chelates of the type [M(a⁀b)$_2$]

Fig. 25

The discussion thus far implies that a particular metal will form either square planar or tetrahedral complexes and this is indeed the general rule. However, there are several exceptions to this and it is well established that the structure of a given complex depends both on the metal and the coordinated ligands. For example, $[Ni(CN)_4]^{2-}$ is square planar whereas $[NiBr_4]^{2-}$ is tetrahedral. It follows that a proper choice of metal and ligand should provide a system in which the energy difference between these two structures is small. Isomers of this type have indeed been recently observed for certain Ni(II) and Co(II) complexes (Fig. 26).

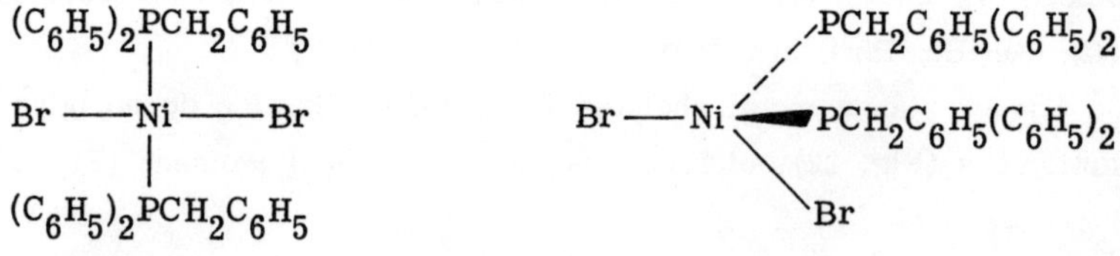

Square - planar

Tetrahedral

Isomers of Ni(II) complexes differing in arrangement of ligands

Fig. 26

b) Six-Coordinated Metal Complexes. The most common coordination number of metals is six, and the stereochemistry of these systems has been studied most extensively for complexes of Co(III). This is because the complexes of Co(III) are stable and slow to react, thus permitting the isolation and characterization of geometrical and optical isomers. The same is true for complexes of Cr(III), Rh(III), Ir(III), Ru(III) and Pt(IV) as well as other metals. In most cases (exceptions are mentioned at the end of this section) six-coordinated metal complexes have an octahedral structure.

This structure permits the existence of a large number of stereoisomers and many of these have been obtained. For example, a complex of the type [Mabcdef] can in theory form thirty stereoisomers, fifteen geometrical isomers, each existing in two enantiomeric forms. No one has as yet obtained all of the isomers of such a complex, but a compound containing six different ligands has been prepared, viz. [$Pt(C_5H_5N)NH_3(NO_2)ClBrI$].

Complexes of the types [Ma_4b_2] and [Ma_3b_3] are very common. These systems exist in the form of two geometrical isomers (Fig. 27).

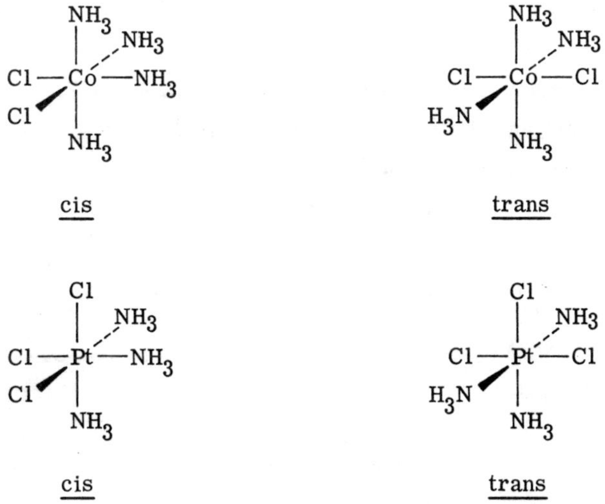

Geometrical isomers of [$Co(NH_3)_4Cl_2$]$^+$ a [Ma_4b_2] complex, and [$Pt(NH_3)_3Cl_3$]$^+$, a [Ma_3b_3] complex

Fig. 27

The cis isomers have adjacent (angles of 90°) like groups, whereas trans isomers have like groups in opposite (angles of 180°) positions. Unsymmetrical bidentate ligands such as a͡b also form cis-trans metal chelates of the type [M(a͡b)$_3$] (Fig. 28).

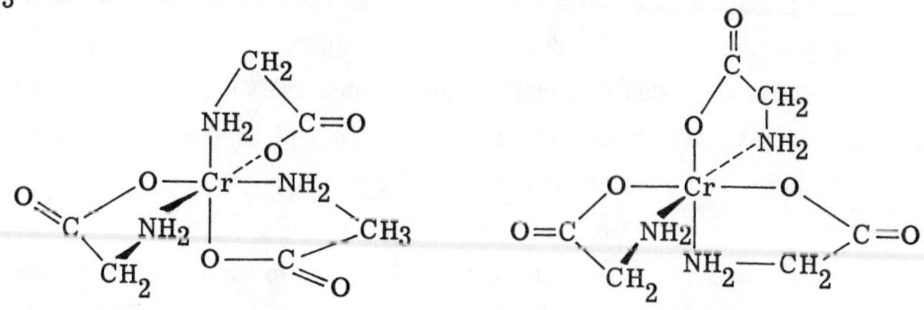

cis   trans

Geometrical isomers of [Cr(NH$_2$CH$_2$COO)$_3$], a [M(a͡b)$_3$] chelate. Each form also has a mirror image isomer.

Fig. 28

Metal chelates of the type [M(a͡a)$_3$] are chiral and give rise to enantiome (Fig. 29). There are many examples of optically active six-coordinated metal complexes. The enantiomers are separated by methods similar to those used for organic compounds and the same experimental techniques are used to detect optic activity in metal complexes as are used in organic compounds.

dextro   levo

Enantiomers of [Co(NH$_2$C$_2$H$_4$NH$_2$)$_3$] $^{3+}$, a [M(a͡a)$_3$] complex.

Note that for simplicity the diamine chelate in the figure does not include C and H

Fig. 29

The dextrorotatory isomer $(+)-[Co(NH_2C_2H_4NH_2)_3]^{3+}$ was examined by means of x-rays and its absolute configuration was determined. This isomer has the structure shown in Fig. 29 for the dextro form. With this structure as a standard it is possible to assign structures to related complexes by comparing their optical rotation at different wavelengths with that of the standard.

There are many complexes of the type $[M(a\,a)_2b_2]$, and these form cis-trans isomers, but only the cis form can exist in enantiomeric forms (Fig. 30).

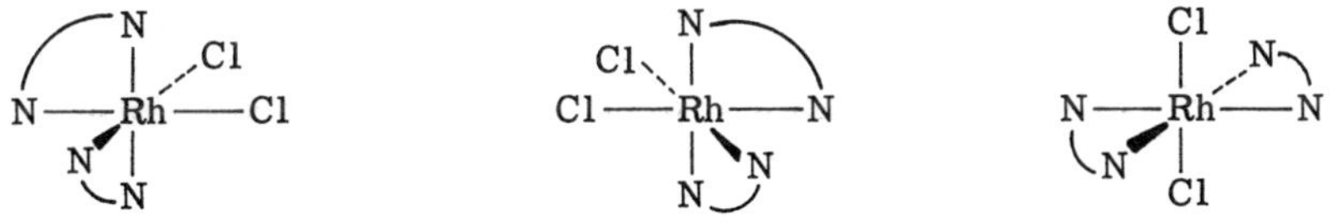

| cis-D | cis-L | trans |

Stereoisomers of $[Rh(NH_2C_2H_4NH_2)Cl_2]^+$, a $[M(a\,a)_2b_2]$ complex

Fig. 30

Fewer complexes such as $[Co(NH_2C_2H_4NH_2)(NH_3)_2Cl_2]^+$ are known, and these exist in the form of three geometrical isomers, one of which is chiral. Mention was made above (Fig. 28) that the cis-trans isomers of $[M(a\,b)_3]$ each form mirror image pairs.

A very versatile chelating agent is the beta-diketone acetylacetone, which is given the symbol acac. In its enol form the anion has a charge of -1 and functions as a bidentate ligand. The coordination number of a metal ion is often twice its oxidation state so that M-acac combinations then result in the formation of molecular compounds. For example, $[Be(acac)_2]$, $[Al(acac)_3]$ and $[Zr(acac)_4]$ are nonionic compounds and most metals form such species. Many of these substances have the property, unusual for metal compounds, of being fairly volatile. Because of this property, it has been said that "the acetylacetonate anion gives wings to metals". Many of these compounds have been resolved by preferential adsorption on optically active quartz or on some sugars (T. Moeller and E. Guylas, 1958). One other point of interest is that these metal chelates have quasiaromatic properties and undergo such reactions as acylation, nitration and halogenation which are characteristic of reactions of aromatic compounds (Fig. 31).

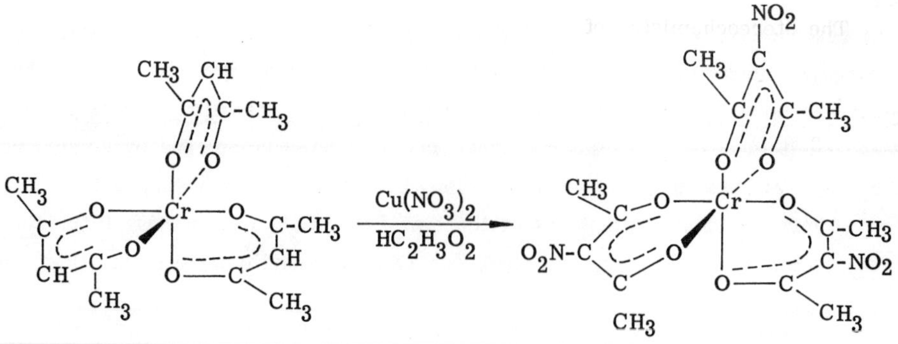

Quasiaromatic reactivity of [Cr(acac)$_3$] (J. P. Collman, 1963)

Fig. 31

Not only do bidentate ligands, a⌢a, form metal complexes, but also tridentate, a⌢a⌢a, quadridentate, a⌢a⌢a⌢a, etc. The most useful chelating agent is ethylenediaminetetraacetate ion which is usually given the symbol EDTA. Because of its geometry, this anion readily coordinates to all six octahedral positions of metal ions. The acid and its sodium salt are produced on a large scale and used in a variety of ways commercially in order to sequester metal ions in solution (S. Charberek and A. E. Martell, "Sequestering Agents", John Wiley and Sons, New York, New York, 1959). Some of the most stable metal chelates are formed by EDTA. Of interest here is the fact that several [M(EDTA)] complexes have been resolved and are found to be optically stable. The enantiomers of the Co(III) chelate are shown in Fig. 32.

dextro   levo

Enantiomers of [Co(EDTA)]$^-$, where EDTA = $(OOCCH_2)_2 NC_2H_4N(CH_2COO)_2^{4-}$

Fig. 32

The stereochemistry of Co(III) complexes of the quadridentate tetramine of $NH_2C_2H_4NHC_2H_4NHC_2H_4NH_2$, given the symbol trien, is currently under active investigation (D. A. Buckingham, P. A. Marzilli and A. M. Sargeson, 1967). Five different stereoisomers have been isolated and characterized for complexes of the type $[Co(trien)a_2]^+$ (Fig. 33). Three of these are geometrical isomers,

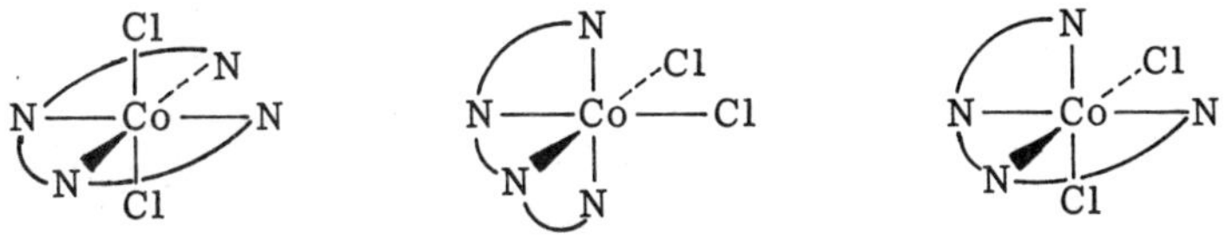

trien α-cis β-cis

Geometrical isomers of $[Co(trien)Cl_2]^+$. The α- and β-cis forms are chiral.

Fig. 33

there being only one trans form but two cis forms. The cis forms differ in that the three chelate rings are all in different planes for the α isomer, whereas two of the chelate rings are in the same plane with the third in a different plane for the β isomer. Each of the cis isomers has been resolved yielding the dextro and levo forms.

In addition to the formation of metal chelates with ligands of different geometries as described above, there has also been an increasing interest in the application of optically active ligands. For example propylenediamine (pn = $NH_2\overset{*}{C}HCH_2NH_2$) has much the same tendency to form metal chelates as
$\underset{|}{\overset{}{}}$
CH_3
does ethylenediamine (en = $NH_2CH_2CH_2NH_2$). The chief difference between the two is that pn contains an asymmetric carbon and thus can be used in the optically active forms of (+)-pn or (−)-pn.

When a dissymmetric metal chelate of the type $[M(a\frown a)_3]$ contains an optically active ligand such as (+)-pn, an asymmetric bias is imparted to the system and the two possible isomers, D*-$[M(+)-pn_3]$ and L*-$[M(+)-pn_3]$ (where D* and L* indicate opposite configuration at the metal), are no longer obtained in equal amounts, since they are now diastereoisomers rather than enantiomers.

This preference for one isomer over the other is called <u>ligand stereospecificity</u>. The specificity may vary from being negligible to being complete. For example, the chelation of Co(III) with (-)-PDTA^{4-} ((-)-propylenediaminetetraacetate ion) yields exclusively L*-[Co(-)-PDTA]$^-$ and none of the D* isomer. This stereo-specific behavior has been used to prepare optically active complexes and also to resolve amines, carboxylic acids and amino acids.

The stereospecificity observed in these systems has been successfully accounted for on the basis of conformational analysis (E. J. Corey and J. C. Bailar, 1959). This analysis is rather involved and will not be discussed here. However, it should be mentioned that ligands such as ethylenediamine form five-membered chelate rings with the metal, and these rings are not planar but have a puckered structure. The groups attached to adjacent carbon and nitrogen atoms are in a staggered arrangement of two possible conformations which are mirror images (Fig. 34). The method of conformational analysis leads to estimates of

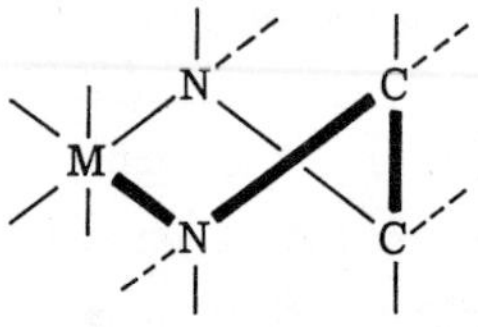

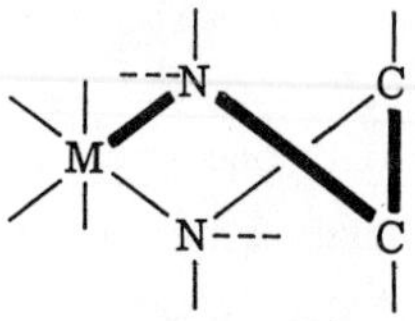

k   k'

The conformations of coordinated ethylenediamine rings where bonds to N and C represents axial (-) and equatorial (----) positions. The structure for k is that found in D*-[Co(en)$_3$]$^{3+}$

Fig. 34

the atomic interactions for all possible conformations and serves to show that some conformations are more stable than others. Where this has been checked experimentally there is good agreement with the calculated relative stabilities.

Extensive studies have been made of isomerization (<u>cis</u> ⇌ <u>trans</u> inter-conversion) and racemization (<u>dextro</u> ⇌ <u>levo</u> interconversion) reactions. The system that has been investigated most is

$$\underline{cis}\text{-}[\,Co(en)_2Cl_2\,]^+ \rightleftharpoons \underline{trans}\text{-}[\,Co(en)_2Cl_2\,]^+$$

$$\text{violet} \qquad\qquad\qquad\qquad \text{green}$$

In water solution the process of isomerization is complicated by the intermediate formation of the aquo complexes $[\,Co(en)_2H_2OCl\,]^{2+}$ and $[\,Co(en)_2(H_2O)_2\,]^{3+}$. In methanol solution it is found that the rate of isomerization of $\underline{cis}\text{-}[\,Co(en)_2Cl_2\,]^+$ $\longrightarrow \underline{trans}\text{-}[\,Co(en)_2Cl_2\,]^+$ is slightly slower than the rate of exchange of Cl^- in the complex with radiochloride ion in solution. It is also found that this rate depends only on the concentration of the complex. These results suggest that isomerization takes place by a dissociative intermolecular process (Fig. 35). If

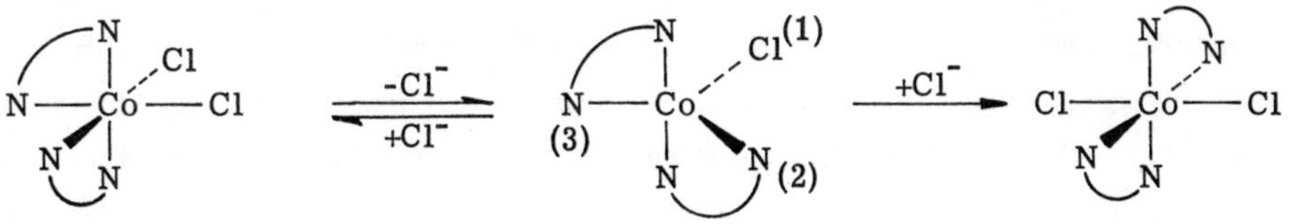

Mechanism of $\underline{cis}$-$\underline{trans}$ isomerization of $[\,Co(en)_2Cl_2\,]^+$

Fig. 35

the Cl^- enters the trigonal bipyramidal intermediate in the 2,3-sector this will yield the $\underline{trans}$ isomer, whereas entry adjacent to Cl in the 1,2- or 1,3-sector returns the $\underline{cis}$ form. Also in support of this intermolecular mechanism is the observation that $(+)$ -$\underline{cis}$-$[\,Co(en)_2Cl_2\,]^+$ loses its optical activity at the same rate at which it exchanges chloride ion.

Optically active complexes of the type $[\,M(a\,\frown\,a)_3\,]$ appear to racemize by means of a chelate ring opening and closing process (Fig. 36). For $[\,Cr(C_2O_4)_3\,]^{3-}$, this mechanism is supported by the observation that the rate of racemization is much faster than the rate of oxalate ion exchange.

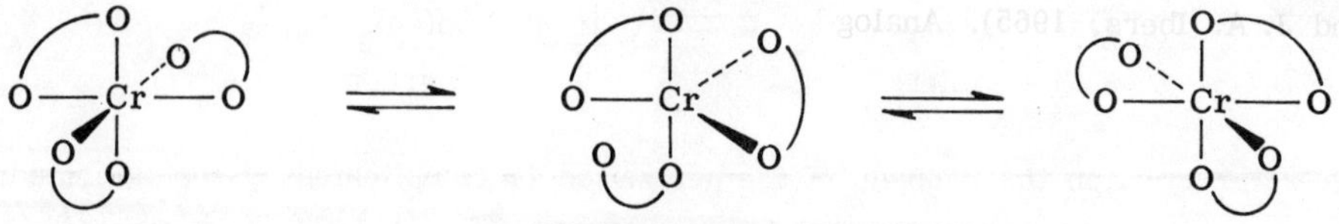

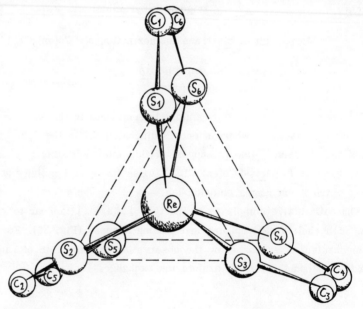

dextro   symmetrical   levo

Mechanism of racemization of $[Cr(C_2O_4)_3]^{3-}$

Fig. 36

Finally, although six-coordinated metal complexes usually have an octahedr
structure, some exceptions to this have recently been discovered. Certain
chelating groups containing sulfur ligand atoms form metal complexes of the type
$[M(a\frown a)_3]$ which have a trigonal prismatic structure. The structure of such a
Re compound was determined by means of X-ray studies (Fig. 37) (R. Eisenberg

Fig. 37   A perspective drawing of the coordination geometry of the molecule

$Re\left(\begin{array}{c} S-C-C_6H_5 \\ \parallel \\ S-C-C_6H_5 \end{array}\right)_3$. The phenyl rings are not shown.

and J. A. Ibers, 1965). Analogous compounds of W, Mo and V appear to have the same structure.

c) Metal Complexes with Coordination Numbers Other than Four or Six. The most common coordination number of metals is six, the next most common is four, but metal complexes are known for all other numbers between two and twelve. The coordination number of two is found in complexes such as $[Ag(NH_3)_2]^+$, $[Au(CN)_2]^-$ and $[Hg(NH_3)_2]^{2+}$, all of which have a linear structure. Planar three-coordinated complexes of the type $[M(PR_3)_2I]$ are known for $M(I) = Cu$, Ag, and Au. There are many examples of five-coordinated systems. Most of these have a trigonal bipyramidal structure, e.g. $Fe(CO)_5$, but some have a tetragonal pyramidal structure, e.g. $[Ni(CN)_5]^{3-}$ (E. L. Muetterties and R. A. Schunn, Quart. Revs. London, 20, 245 (1966)). One specific example for each coordination number greater than six can be given as follows: (7), TaF_7^{2-}; (8), $[Gd(H_2O)_6Cl_2]^+$; (9), $[ReH_9]^{2-}$; (10), $[LaEDTA(H_2O)_4]^-$; (11), $[Th(H_2O)_3(NO_3)_4]$; (12), $[Ce(NO_3)_6]^{3-}$. A variety of different structures are known for these metal complexes with large coordination numbers (E. L. Muetterties and C. M. Wright, Quart. Revs., London, 21, 109 (1967)).

VII. Dynamic Stereochemistry

The discussion so far has dealt principally with the stereochemical aspects of structure (configuration, conformation). There is another important aspect of stereochemistry concerned with the steric course of chemical reactions. This aspect is sometimes called "dynamic stereochemistry". Dynamic stereochemistry is of interest in two quite different contexts. One of these concerns the planned or directed synthesis of a desired stereoisomer, or so-called "stereoselective synthesis". As has already been seen, frequently only one of several possible stereoisomers occurs in nature or is of utility (say as a drug). If this stereoisomer is a pure enantiomer, the method of obtaining it is generally by resolution of a dl-pair by one or other of the methods described earlier, though occasionally asymmetric synthesis (especially enzymatic asymmetric synthesis) may profitably be utilized. More commonly, however, one of several possible diastereoisomers is wanted. For example, the natural product cholesterol has, in addition to an enantiomer, 254 diastereoisomers (127 dl-pairs). In a projected synthesis of cholesterol, to have to separate 128 diastereoisomeric dl-pairs (preceding final resolution) would be a prohibitive task. Clearly, the more rational approach is to direct the synthesis in such a way that only the desired diastereoisomer (plus possibly a very few of the others, if that is unavoidable) is obtained. The stereo-selective synthesis of cis- and trans-4-t-butylcyclohexanol from 4-t-butylcyclo-hexanone shown in Fig. 38 serves as a simple example. Ordinary chemical or catalytic reduction of the ketone gives a mixture of cis- and trans-4-t-butyl-cyclohexanol; if this mixture contains the two diastereoisomers in equal pro-portions, the reaction leading to it is "non-stereoselective".

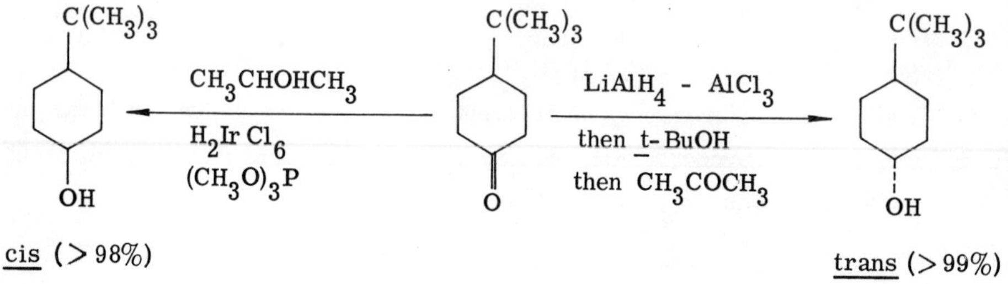

Stereoselective reduction of 4-t-butylcyclohexanone

Fig. 38

A somewhat different aspect of dynamic stereochemistry is concerned with the mechanism of reaction itself. Chemists are interested in how reactions procee and use a variety of tools--such as kinetic study, study of effects of minor struc-tural modifications, isotope effect studies, etc.--to obtain information on the inti-mate details of the course of chemical change. Stereochemical studies are of great importance in helping to elucidate reaction mechanism. As an example of a reaction in whose understanding stereochemical considerations have played an important part, the nucleophilic substitution reaction may be considered in some detail. This is a reaction of the type $R-X + Y:^- \rightarrow R-Y + :X^-$ where R is an organic radical or group and X and Y are bases, either organic or inorganic. Stereochemical information about this reaction was obtained by Walden (1896) who studied the hydrolysis of (+)-chlorosuccinic acid (Fig. 39) with potassium hydroxide

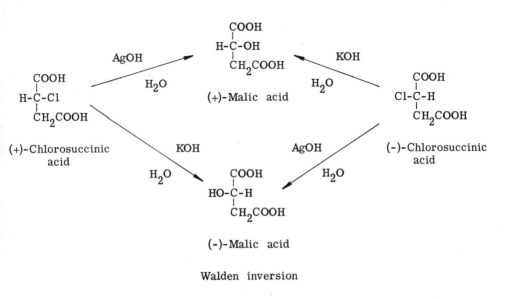

Walden inversion

Fig. 39

and found it to give (-)-malic acid (Fig. 39). However, when the hydrolysis was carried out in the presence of moist silver oxide, (+)-malic acid resulted. It is clear that in either one or the other of these two kinds of hydrolysis, an inversion of configuration must take place (Fig. 39; in this figure the inversion step has been arbitrarily assigned to the KOH reaction and the retention step to the AgOH reaction). The inversion reaction thus discovered has been called the "Walden inversion" but it was not known to Walden in which of the two types of hydrolysis (KOH or AgOH) inversion occurred; and in fact it took almost 30 years before the stereochemical course of any displacement reaction was elucidated. It is now known that the uncommon step in the Walden inversion scheme is the step involving retention of configuration; inversion of configuration is the common course of nucleophilic displacement. The transition state or activated complex in at least one type of nucleophilic displacement, the so-called bimolecular type, is shown in Fig. 40; it is clear, from the diagram, why this step involves inversion. (The incoming nucleophile, here CH_3COO^-, approaches the molecule from the back while the leaving group Br^- recedes in front.)

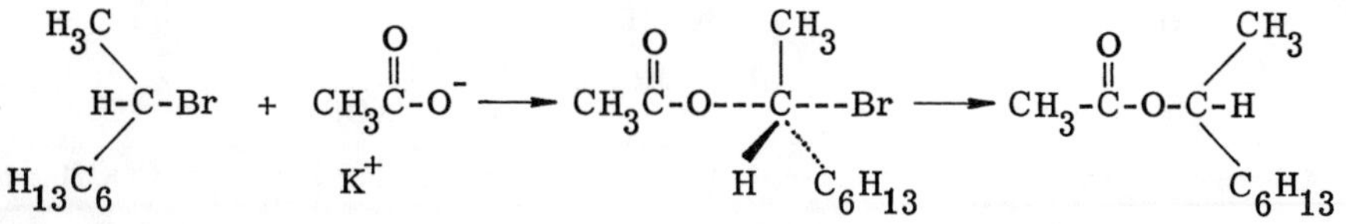

$$\underset{\substack{\text{H}_3\text{C} \\ \text{H-C-Br} \\ \text{H}_{13}\text{C}_6}}{} + \underset{\substack{\text{K}^+}}{\text{CH}_3\overset{\text{O}}{\overset{\|}{\text{C}}}\text{-O}^-} \longrightarrow \text{CH}_3\overset{\text{O}}{\overset{\|}{\text{C}}}\text{-O---}\underset{\substack{\text{H} \quad \text{C}_6\text{H}_{13}}}{\overset{\text{CH}_3}{\text{C}}}\text{---Br} \longrightarrow \text{CH}_3\text{-}\overset{\text{O}}{\overset{\|}{\text{C}}}\text{-O-}\underset{\substack{\text{C}_6\text{H}_{13}}}{\overset{\text{CH}_3}{\text{C}}}\text{-H}$$

(S)-(+)-2-Bromooctane   Activated complex   (R)-(-)-2-Octyl acetate
                                                        + K$^+$ Br$^-$

Inversion in bimolecular nucleophilic substitution

Fig. 40

The stereochemical course of numerous other reactions is known. Among those involving saturated (quadriligant) carbon are a number of rearrangement reactions, exemplified by the so-called Hofmann rearrangement shown in Fig. 41. In this rearrangement and many similar ones, the migrating group (here $C_6H_5CHCH_3$) retains its configuration.

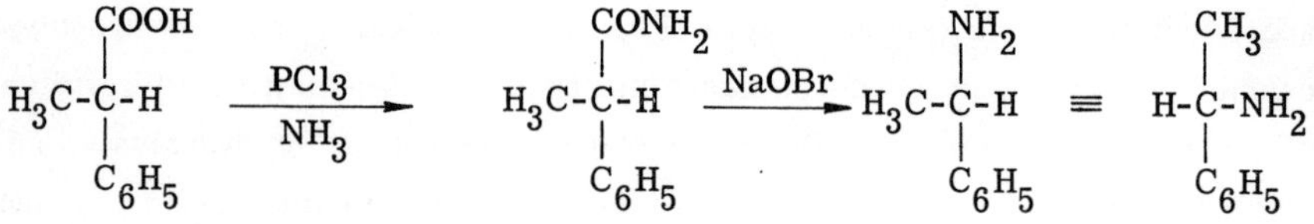

$$\underset{\substack{\text{C}_6\text{H}_5}}{\overset{\text{COOH}}{\text{H}_3\text{C-C-H}}} \xrightarrow[\text{NH}_3]{\text{PCl}_3} \underset{\substack{\text{C}_6\text{H}_5}}{\overset{\text{CONH}_2}{\text{H}_3\text{C-C-H}}} \xrightarrow{\text{NaOBr}} \underset{\substack{\text{C}_6\text{H}_5}}{\overset{\text{NH}_2}{\text{H}_3\text{C-C-H}}} \equiv \underset{\substack{\text{C}_6\text{H}_5}}{\overset{\text{CH}_3}{\text{H-C-NH}_2}}$$

(S)-(+)-Hydratropic   Hofmann rearrangement   (S)-(-)-α-Phenethylamine
acid

Fig. 41

Among reactions of unsaturated compounds, addition to olefins and acetylene has been extensively studied. Catalytic reduction of acetylenes usually gives main cis-olefins (Fig. 42) because the hydrogen from the catalyst surface approaches the triple bond from one side. Similarly, hydroboration with B_2H_6 followed by acetic acid hydrolysis leads to cis-olefin (Fig. 42); the first step involves simultaneous addition of B and H from a single B_2H_6 molecule and logically proceeds in cis fashion, the second step may involve a carbanion intermediate (which is subsequently protonated); it is known that unsaturated carbanions, abC=Cd:$^-$, similar to oximes (see above) maintain their geometric configuration. In contrast

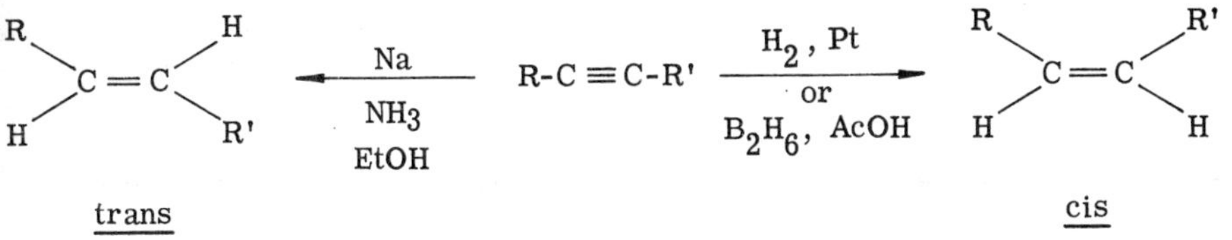

Stereochemistry of reduction of acetylenes to olefins

Fig. 42

reduction of acetylenes with sodium and ammonia involves overall <u>trans</u> addition of hydrogen to give a <u>trans</u>-olefin; this reaction may well involve a dianion, $R-\bar{C}=\bar{C}-R'$ or anion-radical $R-\dot{C}=\bar{C}-R'$ intermediate in which the tendency of the electrons to maintain maximum distance (because of electrostatic repulsion) is responsible for the stereochemical outcome. Addition of bases, such as $CH_3O^-Na^+$ or $CH_3S^-Na^+$ to acetylenes also proceeds in <u>trans</u> fashion (to give <u>cis</u> olefin!) as does electrophilic addition.

Catalytic hydrogenation of olefins involves predominantly <u>cis</u> (or <u>syn</u>) addition, similarly as in the case of acetylenes. Reactions involving simultaneous additions of two parts of one and the same molecule also follow predominantly a <u>cis</u> course. Among such reactions are the formation of glycols RCHOHCHOHR' from olefins RCH=CHR' by oxidation with osmium tetroxide or potassium permanganate and the formation of alcohols, RR'CHCHOHR'' from olefins RR'C=CHR'' by hydroboration (treatment with B_2H_6) followed by oxidation with basic hydrogen peroxide.

A rather common type of addition of olefins is the so-called electrophilic addition in which an electron-poor part of the reagent adds to the electron-rich double bond to give a positively charged intermediate which then reacts with a base to give the final product. In a number of cases which have been studied (though by no means universally) the intermediate is a cyclic one and the overall course of reaction is one of <u>trans</u> (or <u>anti</u>) addition, as shown in Fig. 43. Reagents which add in this fashion are the halogens, 2,4-dinitrobenzenesulfenyl chloride, hypochlorous acid, formaldehyde and acid (Prins reaction) among others. The <u>trans</u> addition of peracetic acid in the presence of mineral acid is of special interest because the cyclic intermediate (epoxide) is a stable compound and can

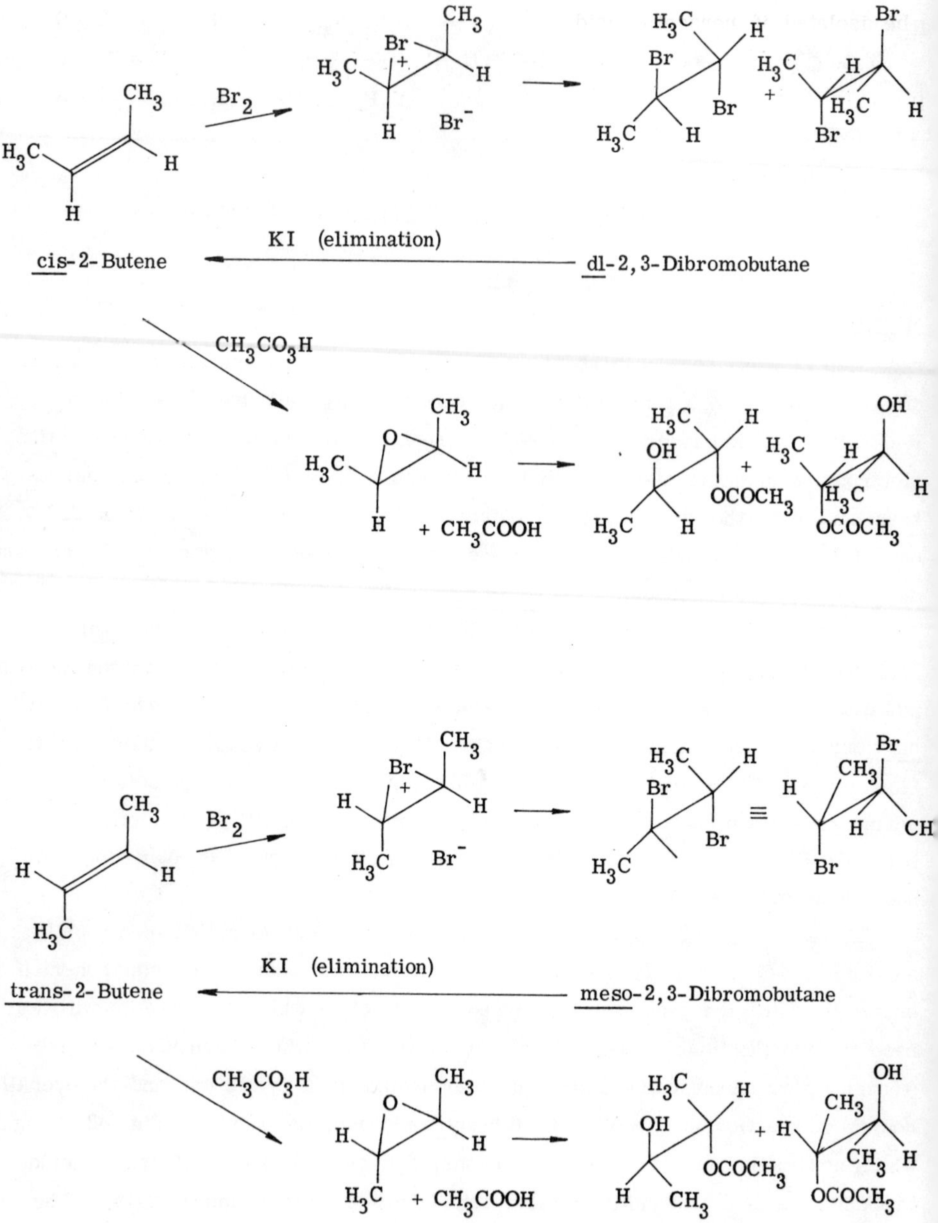
Br2
KI (elimination)
cis-2-Butene
dl-2,3-Dibromobutane
CH3CO3H
+ CH3COOH
Br2
KI (elimination)
trans-2-Butene
meso-2,3-Dibromobutane
CH3CO3H
+ CH3COOH

Electrophilic trans addition and trans elimination

Fig. 43

be isolated if peracetic acid is added to the olefin in the absence of mineral acid.

It is worth mentioning here that cyclic intermediates similar to the bromonium ion generated in the addition of bromine to olefin also occur in certain substitution reactions, for example the reaction of the active form of threo-3-bromo-2-butanol with hydrogen bromide which is shown, in Fig. 44, to give rise to dl-2,3-dibromobutane via a cyclic ion. Formation of such a cyclic

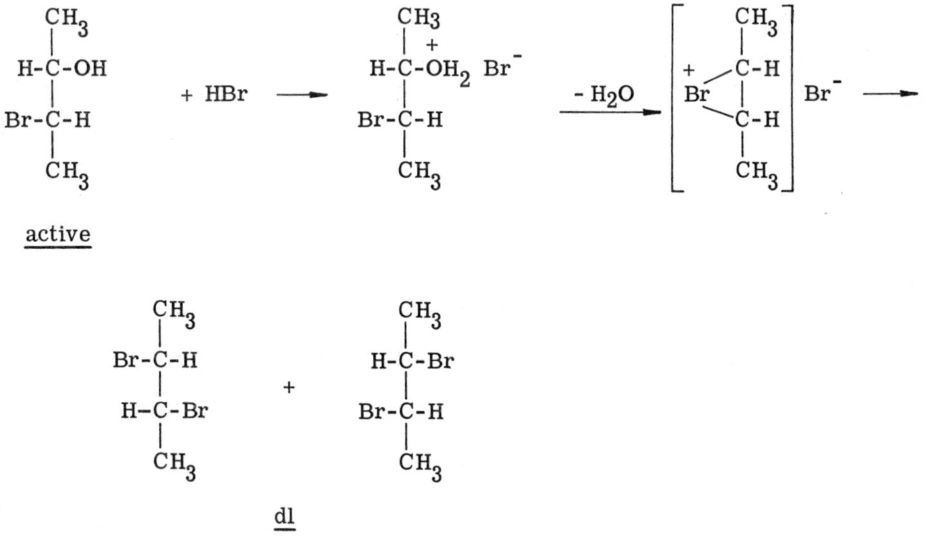

$$
\begin{array}{c}
CH_3 \\
| \\
H-C-OH \\
| \\
Br-C-H \\
| \\
CH_3 \\
\text{active}
\end{array}
\quad + HBr \longrightarrow
\begin{array}{c}
CH_3 \\
| \ + \\
H-C-OH_2 \ Br^- \\
| \\
Br-C-H \\
| \\
CH_3
\end{array}
\quad - H_2O \longrightarrow
\left[
\begin{array}{c}
CH_3 \\
| \\
+ \ C-H \\
Br \ | \\
C-H \\
| \\
CH_3
\end{array}
\right] Br^- \longrightarrow
$$

$$
\begin{array}{c}
CH_3 \\
| \\
Br-C-H \\
| \\
H-C-Br \\
| \\
CH_3
\end{array}
\quad + \quad
\begin{array}{c}
CH_3 \\
| \\
H-C-Br \\
| \\
Br-C-H \\
| \\
CH_3
\end{array}
$$

dl

Reaction of threo-3-bromo-2-butanol with hydrogen bromide

Fig. 44

ion in nucleophilic substitution requires the presence of an active ("participating") group in the vicinity of the reactive site and is therefore said to be a reaction involving "neighboring group participation". It should be noted that, contrary to ordinary nucleophilic substitution, substitution with neighboring group participation involves retention rather than inversion of configuration. Neighboring group participation of the COOH group may well be responsible for the retention of configuration involved in the reaction of chlorosuccinic acid with AgOH (Fig. 39).

Addition to olefins may also proceed _via_ free radical intermediates, via anionic intermediates or via diradical intermediates of the type :Cab; the stereo-chemistry of such addition has been studied but is not as straightforward as that in the addition reactions already discussed.

Among the many other reactions whose stereochemical course has been elucidated only the elimination reaction will be mentioned here. As shown in Fig. 43, just as addition of bromine to 2-butene follows a _trans_-stereochemical course, so does the reverse reaction, ionic elimination of bromine from 2,3-di-bromobutane to give 2-butene. In contrast, the molecular elimination of dimethyl hydroxylamine from the N-oxide of the 2-dimethylamino-3-phenylbutanes takes a _cis_ course (Fig. 45).

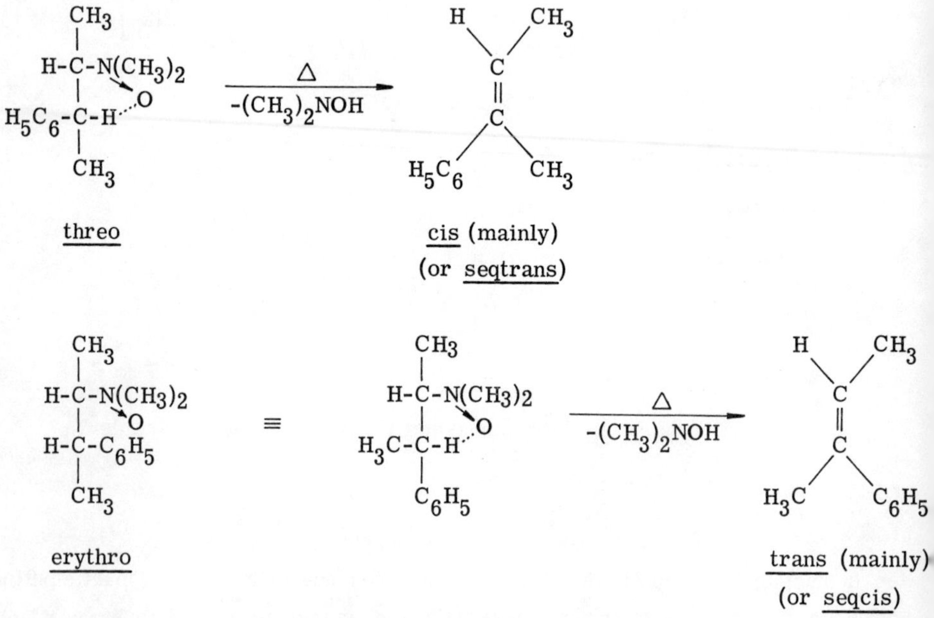

Molecular _cis_-elimination

Fig. 45

VIII. Stereoregular Polymerization

Polymers are very large molecules with recurring small groups or units of atoms generally derived from smaller molecules ("monomers") which are the precursors of the polymers and from which the polymers are obtained by a process of chemical stringing up or "polymerization". Thus a protein is a polymer of α-aminoacid units, cellulose is a polymer of glucose units, etc. The monomers may be likened to paper clips which are hooked together in a large chain to make the analog of a polymer.

Among typical synthetic polymers are those derived from so-called "α-olefins", $R-CH=CH_2$, the structure of the polymer being $-(\underset{R}{CH}-CH_2)_n^-$.

Representative examples are the acrylonitrile polymer Orlon (R = CN) and polystyrene, a hard plastic used in insulators and in certain household articles $(R = C_6H_5)$.

In 1955, G. Natta and coworkers in Milan, Italy, examined polymers derived from propylene, $CH_3CH=CH_2$ and having the general formula $-(\underset{CH_3}{CH}-CH_2)_n^-$ by the technique of X-ray diffraction. They found that a few of the polymers investigated (specifically those obtained from the monomer by means of an aluminum trialkyl--titanium chloride or so-called "Ziegler" catalyst--named after its discoverer Karl Ziegler in Muelheim, Germany--or by means of a chromium trioxide--silica-alumina catalyst) showed a much higher regularity in their structure than other polypropylene polymers. Extensive investigation disclosed that polymers of α-olefins may have the stereochemically distinct structures shown (by means of Fischer projection formulas) in Fig. 46. The arrangement in which all the α-groups are on the same side of the polymer chain is called "isotactic". The arrangement in which the groups alternate is called "syndiotactic". A random type arrangement is called "atactic" or "heterotactic". (There are also polymers having blocks of stereoregularly placed monomer units followed by other blocks of a different but also regular arrangement. Such polymers are called "stereoblock polymers".)

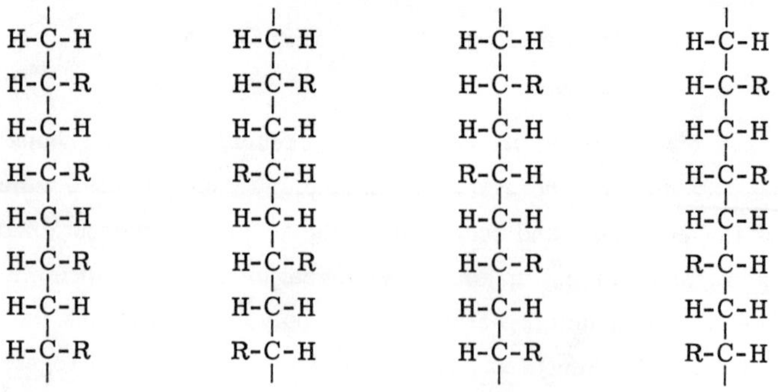

Isotactic	Syndiotactic	Atactic or heterotactic	Stereoblock
H-C-H	H-C-H	H-C-H	H-C-H
H-C-R	H-C-R	H-C-R	H-C-R
H-C-H	H-C-H	H-C-H	H-C-H
H-C-R	R-C-H	R-C-H	H-C-R
H-C-H	H-C-H	H-C-H	H-C-H
H-C-R	H-C-R	H-C-R	R-C-H
H-C-H	H-C-H	H-C-H	H-C-H
H-C-R	R-C-H	H-C-R	R-C-H

Stereoregular polymers derived from α-olefins

Fig. 46

The effect of stereoregularity on physical properties is striking. For example, whereas ordinary polypropylene (atactic) is a soft rubbery material, the isotactic variety is a fiber-forming polymer which may be spun and woven into a cloth. It is therefore not surprising that Natta and Ziegler received the Nobel prize (1963) for their discoveries of stereoregular polymers and the catalyst systems needed to make them.

Important stereoregular polymers are found in nature. The proteins are polymers of amino acids all having the Fischer projection formula shown in Fig. 47 (they belong to the so-called "L-series" because the NH_2-group is on the left). Rubber and gutta-percha (Fig. 47) are naturally occurring polymers of isoprene (Fig. 47) in which the stereoregularity is of a somewhat different type: rubber is all-cis whereas gutta-percha is all-trans about the double bonds. Correspondingly there is a vast difference in properties, only rubber being elastic and resilient whereas gutta-percha is a flexible, soft but not resilient plastic material. Methods are known to synthesize pure (all-cis) rubber from its monomer, isoprene, on an industrial scale, using appropriate catalysts.

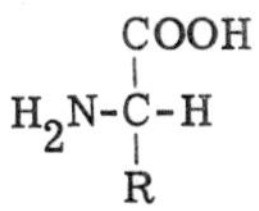

COOH
|
H_2N-C-H
|
R

L-Amino acid

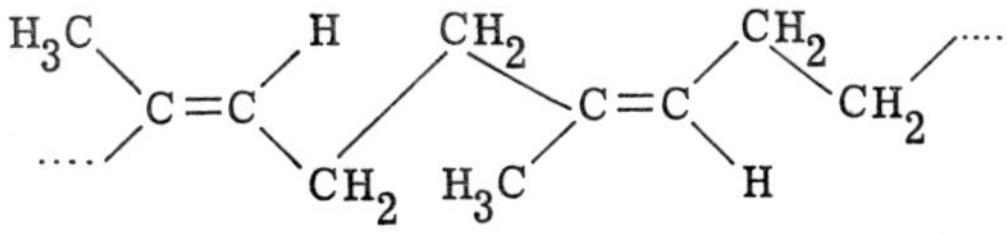

$CH_2=C-CH=CH_2$
|
CH_3

Isoprene

Rubber (cis chain)

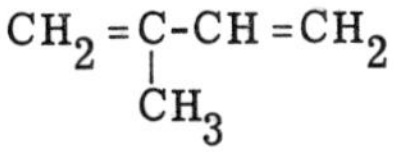

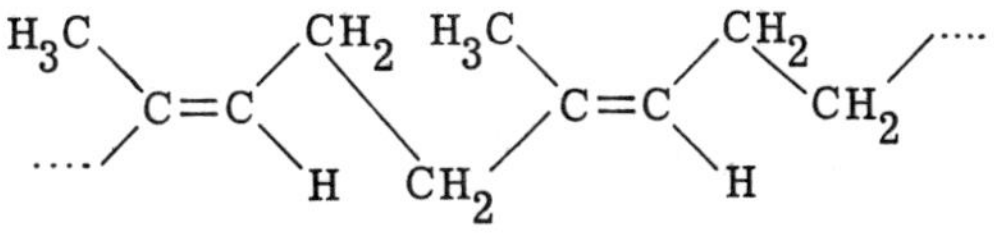

Gutta-percha (trans chain)

Naturally occurring stereoregular polymers and their monomers

Fig. 47

IX.   General References

General Stereochemistry:

E. L. Eliel, Stereochemistry of Carbon Compounds, McGraw-Hill Book Co.,
Inc., New York, 1962.

K. Mislow, Introduction to Stereochemistry, W. A. Benjamin, Inc., New York,
1965.

N. L. Allinger and E. L. Eliel, editors, Topics in Stereochemistry, Interscience
Division, John Wiley & Sons, Inc., New York; Vols. 1,2, 1967, Vol. 3, 1968.

Inorganic Stereochemistry:

J. C. Bailar, Jr., editor, The Chemistry of Coordination Compounds, Reinhold
Publishing Co., New York, 1956.

F. Basolo and R. C. Johnson, Coordination Chemistry, W. A. Benjamin, Inc.,
New York, 1964.

F. P. Dwyer and D. P. Mellor, editors, Chelating Agents and Metal Chelates,
Academic Press, Inc., New York, 1964.

Conformational Analysis:

E. L. Eliel, N. L. Allinger, S. J. Angyal and G. A. Morrison, Conformational
Analysis, Interscience Division, John Wiley & Sons, New York, 1965.

M. Hanack, Conformation Theory, Academic Press, Inc., New York, 1965.

J. McKenna, Conformational Analysis of Organic Compounds, Lecture Series
No. 1, The Royal Institute of Chemistry, London, 1966.

X. PROBLEMS

1) (pp. 1-3) Define the terms "constitution" and "configuration". Do
   the following pairs differ from each other in constitution or in configuration?

   a) Lactic acid, $CH_3CHOHCOOH$ (of unspecified rotation) and
       β-Hydroxypropionic acid, $HOCH_2CH_2COOH$.
   b) (+)-Lactic acid and (-)-Lactic acid.
   c) (-)-Lactic acid and β-Hydroxypropionic acid.
   d) 3-Methylcyclohexanol and 4-methylcyclohexanol.
   e) cis- and trans-3-Methylcyclohexanol.
   f) (+)-cis-3-Methylcyclohexanol and (-)-cis-3-methylcyclohexanol.
   g) cis- and trans-4-Methylcyclohexanol.
   h) 1-Chloropropene, $ClCH=CHCH_3$, 2-chloropropene, $CH_2=CClCH_3$
       and 3-chloropropene (allyl chloride), $CH_2=CHCH_2Cl$.
   i) cis-1-Chloropropene and trans-1-chloropropene.

2) (p. 3) What are stereoisomers? Do they differ in constitution?

3) (p. 5) Mention one important experimental and one important con-
   ceptual discovery of Louis Pasteur's.

4) (p. 6) Define: a) chiral center, b) enantiomer, c) racemic modification. Give an example for each.

5) (p. 7) Give an example for a chiral and for an achiral compound. Is
   the chiral compound you have chosen asymmetric? If it is, give
   an example of a chiral compound which is not asymmetric.

6) (p. 8) Are the following pairs enantiomers or diastereoisomers?

   a) (+)-Tartaric acid and (-)-tartaric acid. (Formulas on p. 5).
   b) (-)-Tartaric acid and meso-tartaric acid. (Formulas on p. 5).
   c) cis- and trans-1, 2-Dichloroethylene, $ClCH=CHCl$.
   d) (+)- and (-)-cis-3-Methylcyclohexanol.
   e) cis- and trans-3-Methylcyclohexanol.
   f) Crystalline (-)-tartaric acid and crystalline racemic tartaric acid.

7) (p. 8) Do enantiomers differ between each other in the following properties? a) Boiling point. b) Melting point. c) Infrared spectrum. d) Nmr spectrum. e) Ultraviolet spectrum. f) Optical rotation. g) Optical rotatory dispersion or circular dichroism. h) Refractive index. i) Dipole moment. j) Free energy. k) Reactivity toward an achiral chemical reagent. l) Reactivity toward a chiral chemical reagent, such as an enzyme.

8) (p. 9) Do diastereoisomers differ in the above properties (question 7)?

9) (p. 9) For compounds of the same constitution, are there some

    a) which display enantiomerism but no diastereoisomerism?

    b) which display diastereoisomerism but no enantiomerism?

    c) which display both diastereoisomerism and enantiomerism?

Give examples.

10) (p. 10) The term "cis-trans Isomerism" (formerly "geometrical isomerism") is often used in the literature. Does it correspond to enantiomerism or diastereoisomerism?

11) (p. 11) Write Fischer projections for each of the following:

    a) Bromochloromethanesulfonic acid (enantiomer of your choice).

    b) meso-2, 3-Dibromobutane

    c) Active 2, 3-dibromobutane (enantiomer of your choice).

    d) erythro-2, 3-Pentanediol (enantiomer of your choice).

    e) threo-2, 3-Pentanediol (enantiomer of your choice).

12) For the compound given in question 11b, write all three Newman projection formulas.

13) For the compounds you have drawn in answer to question 11, parts c)- e), write one Newman projection formula and the corresponding saw-horse formula. Make sure you depict the same configuration you have chosen in answering question 11).

14) (pp. 12-14) Indicate whether the following are (R)- or (S)-.

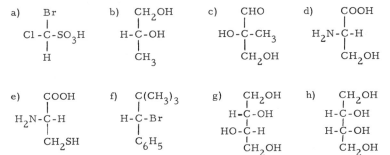

a)   Br
         |
   $Cl-C-SO_3H$
         |
        H

b) CH_2OH
        |
   $H-C-OH$
        |
      CH_3

c)   CHO
         |
  $HO-C-CH_3$
         |
    CH_2OH

d)   COOH
         |
  H_2N-C-H
         |
    CH_2OH

e)   COOH
         |
  H_2N-C-H
         |
   CH_2SH

f) $C(CH_3)_3$
        |
  $H-C-Br$
        |
    C_6H_5

g) CH_2OH
        |
  $H-C-OH$
        |
  $HO-C-H$
        |
   CH_2OH

h) CH_2OH
        |
  $H-C-OH$
        |
  $H-C-OH$
        |
   CH_2OH

15) Write Fischer projection formulas for:

   a) (R)-Lactic acid (2-hydroxypropionic acid)

   b) (S)-Alanine (2-aminopropionic acid)

   c) (R)-Phenylmethylcarbinol

   d) (S)-Ethanol-1-$\underline{d}$

   e) (R)-$C_6H_5CHOHC_6H_4$-Cl-$\underline{p}$

   f) (R, R)-Tartaric acid

   g) (S)-3-Methyl-1-pentene

   h) (R, R)-1, 3-Cyclohexanediol

16) (pp. 14-16) For each of the following, indicate the total number of stereoisomers, the number of $\underline{dl}$-pairs (i.e. half the number of enantiomers) and the number of me$\underline{so}$ or inactive isomers.

   a) Ephedrine, $C_6H_5-CHOH-CH(NHCH_3)-CH_3$

   b) The aldopentoses, $CH_2OH-CHOH-CHOH-CHOH-CHO$

   c) 2-Decalol,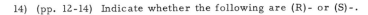

   d) Hydrobenzoin, $C_6H_5-CHOH-CHOH-C_6H_5$

   e) $CH_3-CHCl-CHCl-CHCl-CH_3$

   f) 3-Methylcyclohexanol

g) 4-Methylcyclohexanol

h) 1, 3-Dimethylcyclohexane

i) 1, 3, 5-Trimethylcyclohexane

17) (pp. 16-20) Is it proper to speak of an individual molecule as being racemic? Explain. What is a racemic modification?

18) What is the difference in enthalpy between a pure enantiomer (in dilute solution in an ideal solvent) and a racemic modification (similarly dissolved)? What is the difference in entropy? In free energy? Is racemization an energetically favorable process?

19) (pp. 16-20) Does the addition of hydrogen cyanide to benzaldehyde followed by acid-catalyzed hydrolysis give optically active mandelic acid? Explain. Would bromination of phenylacetic acid, C_6H_5-CH_2-COOH followed by hydrolysis give optically active mandelic acid, C_6H_5-CHOH-COOH? Explain how optically active mandelic acid can be obtained in the laboratory if not by the above methods.

20) (p. 19) When hog kidney acylase is used to resolve an amino acid, the amino group of the racemic acid is first acylated, the acylamino acid is then selectively hydrolyzed in the presence of the enzyme, and the resolved amino acid is separated from the acyl-derivative of its enantiomer. It turns out that the resolved amino acid so obtained is always the natural (generally S) enantiomer, and the residual acylamino acid corresponds to the unnatural (generally R) isomer. Explain why this is so.

21) (pp. 20-22) For each of the following compounds, indicate whether the underlined atoms or groups are equivalent ("homotopic"), enantiotopic or diastereotopic:

a)
```
      COOH
       |
  H-C-H
       |
      COOH
```

b)
```
      COOH
       |
  H₃C-C-CH₃
       |
       H
```

c)
```
      COOH
       |
  H-C-H
       |
  H-C-H
       |
      COOH
```

d) $CH_3CH_2CH_2CH_2CH_3$

e) $CH_3CH_2CH_2CH_2CH_3$

f) The carbinol (CHOH) hydrogens in mesotartaric acid (Fig. 2, p. 5)

g) The carbinol (CHOH) hydrogens in (-)-tartaric acid (Fig 2, p. 5)

h)
$$
\begin{array}{c}
\text{COOH} \\
| \\
\text{H-C-H} \\
| \\
\text{H-C-OH} \\
| \\
\text{COOH}
\end{array}
$$

i)
$$
\begin{array}{c}
\text{COOH} \\
| \\
\text{H-C-H} \\
| \\
\text{HO-C-COOH} \\
| \\
\text{H-C-H} \\
| \\
\text{COOH}
\end{array}
$$

j)
$$
\begin{array}{c}
\text{CH}_2\text{OH} \\
| \\
(\text{CHOH})_n \\
| \\
\text{CH}_3
\end{array}
$$

k) The hydrogen atoms at C-2 and at C-4 in (-)-trihydroxyglutaric acid (Fig. 8, p. 15).

l) The hydrogen atoms at C-2 and at C-4 in either of the meso forms of trihydroxyglutaric acid (Fig. 8, p. 15).

22) For each of the compounds referred to in Question 21, indicate a) whether the protons or groups underlined will, in principle, give rise to identical or separate nmr signals; b) whether they will or will not, in principle, be equivalent toward an enzyme.

23) Indicate whether the two faces of the double bond (C=O or C=C) in each of the following compounds are equivalent ("homotopic"), enantiotopic or diastereotopic:

a) $C_6H_5CH=O$ 　　b) $CH_3\overset{O}{\overset{\|}{C}}CH_3$ 　　c) $CH_3\overset{O}{\overset{\|}{C}}CH(CH_3)C_2H_5$

d) $CH_3\overset{O}{\overset{\|}{C}}CH_2CH_2\overset{O}{\overset{\|}{C}}CH_3$ 　　e) Maleic acid (Fig. 12, p. 29)

f) Fumaric acid (Fig. 12, p. 29).

24) Indicate whether addition of HCN in the case of the carbonyl compounds in Question 23 (a-d) and catalytic addition of deuterium, D_2 (assumed to be cis-addition) in the case of the olefinic acids (23e, f) to the two faces of the double bond will give rise to the same product, to enantiomeric products or to diastereoisomeric products.

25) (pp. 23-27) Consider rotation about the central bond in butane, $CH_3CH_2\text{-}CH_2CH_3$. How many eclipsed conformations are there? How many staggered ones? Which of the staggered ones are gauche? Which anti?

26) The dipole moment of gaseous 1,2-dichloroethane, $ClCH_2CH_2Cl$ is 1.12D (Debye units) at room temperature. What, qualitatively, may you conclude about its conformation?

27) 1,2-Dibromoethane shows a rather simple infrared spectrum in the solid (crystalline) state. When the material is melted, the absorption bands shown by the solid diminish in intensity but do not disappear. A large number of additional bands appear at the same time. Explain.

28) Intramolecular hydrogen bonding (as evidenced by boiling point, infrared spectrum, etc.) is more intense in dl-2,3-butanediol than in the meso isomer. Explain.

29) (pp. 28-31) Would you expect cis-trans isomerism to exist in the following cases:
   a) ClCH=CHCl   b) HOOCCH=CHCOOH   c) The anhydride of b.
   d) $CH_3CH=CHCl$ e) $(CH_3)_2C=CHCH_3$ f) CHCl=C=CHCl
   g) $CH_3CH=C=C=CHCH_3$ h) $CH_2=C=C=CH_2$

30) For the case of $CH_3CH=CHCH_3$, CHCl=CHCl and $CH_3CH=CHCl$, indicate whether the cis or trans isomer has the higher dipole moment, the higher boiling point and the greater stability.

31) (pp. 32-36) Define small, common, medium and large rings. How do these four types differ in ring strain?

32) Compare the ease of formation of a cyclic oxide (oxirane or oxetane, as the case may be) from $ClCH_2CH_2OH$, $ClCH_2C(Me)_2OH$, $ClCH_2CH_2CH_2OH$ and $ClCH_2C(Me)_2CH_2OH$.

33) When an olefin is hydrogenated to the corresponding saturated hydrocarbon, heat is given off. Will more heat be given off in the hydrogenation of cis-2-butene or trans-2-butene? Will the hydrogenation of cis-cyclononene or trans-cyclononene be more exothermic?

34) (p. 36) Give examples of an optically active allene, alkylidene-cycloalkane, spirane.

35) (pp. 36-39) Would you expect the following biphenyl to be capable of resolution?
   a) 2,2'-dicarboxybiphenyl
   b) 2,2'-dicarboxy-6,6'-dibromobiphenyl
   c) 2,6-dicarboxy-2',6'-dibromobiphenyl

Numbering system in biphenyl:

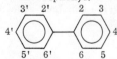

d) 2, 2'-diiodobiphenyl-4-carboxylic acid

e) 2, 2', 6, 6'-tetramethylbiphenyl-3, 3'-dicarboxylic acid

f) 2, 2', 6, 6'-tetramethylbiphenyl-4, 4'-dicarboxylic acid

36) (pp. 39-41) Would you expect to find separable stereoisomers in

a) 1, 4-dichlorocyclohexane   b) chlorocyclohexane (equatorial, axial)

c) $H_3C-C-C_2H_5$
         ‖
         N-OH

d) H_3C-C-O^-
         ‖
         N-CH_3

e) $H_3C-CH-C_2H_5$
        |
        C_3H_7

f) $H_3C-\overset{..}{N}-C_2H_5$
        |
        C_3H_7

g) 1, 2-dimethylcyclopropane   h) N, 2-dimethylethylene imine

37) (p. 41) Can _trans_-cyclooctene be resolved and, if so, how? Would the same type of resolution work for _trans_-cyclononene?

38) (pp. 41-44) Give examples for:

a) An ansa compound   b) A paracyclophane

c) A catenane   d) A helical compound

39) In each of the following cases, optical activity is lost. Explain why.

a) Active $H_5C_2\overset{+}{-}\underset{\underset{C_6H_5}{|}}{\overset{\overset{CH_3}{|}}{N}}-CH_2C_6H_5$ OH^- is reduced, by sodium amalgam,

    to methylethylaniline (and toluene).

b) Active $C_6H_5-CO-CH(CH_3)C_2H_5$ is treated with base.

c) Active $C_6H_5-CHD-CH_3$ is brominated with N-bromosuccinimide the major product being $C_6H_5-CBrD-CH_3$ (through operation of an isotope effect).

d) Optically active $C_6H_5-\underset{\underset{Cl}{|}}{\overset{\overset{CH_3}{|}}{C}}-C_6H_4-OCH_3$-_p_ is solvolyzed, in methyl

    alcohol, to the corresponding methyl ether, $C_6H_5-\underset{\underset{OCH_3}{|}}{\overset{\overset{CH_3}{|}}{C}}-C_6H_4-OCH_3$-_p_.

40) (pp. 47-54) Diagram all possible stereoisomers of each of the following:

    a) Tetrahedral Structures:

        i) $[Be(CH_3COCHCOC_6H_5)_2]$ ii) $[Co(NO)(CO)P(C_6H_5)_3(CN)]^-$

    b) Square-Planar Structures:

        iii) $[Pd(NH_3)_2Cl_2]$ iv) $[Ni(H_2NCH_2COO)_2]$ v) $[Pt(NH_3)(NO_2)BrCl]^-$

    c) Octahedral Structures:

        vi) $[Co(NH_3)_3(NO_2)_3]$ vii) $[Cr(H_2NCH_2CH_2NH_2)_2Cl_2]^+$

        viii) $[Pt(NH_3)_2Br_2Cl_2]$ ix) $[Rh(H_2NCH_2COO)_2(SCN)_2]^-$

41) (pp. 51-58) The complex $[Co(NH_3)_2(NO_2)_4]^-$ reacts with one equivalent of the bidentate ethylene diamine, $H_2NCH_2CH_2NH_2$ ("en") to form $[Co(en)(NH_3)_2(NO_2)_2]^+$ as two isomers, one of which is resolvable. Assuming that no rearrangement takes place during the reaction and noting that en can only span cis positions, designate the structure of the original complex. Justify your answer by means of appropriate diagrams.

42) (pp. 56-58) The rate of racemization of $[Cr(OOCCOO)_3]^{3-}$ is faster than its rate of exchange with radioactive oxalate ion. Also, all twelve of the oxygen atoms in the complex exchange with oxygen-18 of the solvent water (H_2O^{18}) at a rate similar to that for racemization. Explain what these observations tell us about the mechanism of race-mization of this chelate complex.

43) (p. 60) Reduction of dl-α-methylaminopropiophenone, $C_6H_5-CO-CH(NHCH_3)-CH_3$ with sodium amalgam gives a mixture of the diastereoisomeric alcohols, $C_6H_5-CHOH-CH(NHCH_3)-CH_3$. (The erythro isomer is the racemic form of the natural alkaloid (-)-ephedrine.) The threo isomer, called pseudoephedrine, predominates slightly in the mixture. In con-trast, reduction of the same ketone, in the form of the hydrochloride, by hydrogen over a platinum catalyst gives almost exclusively dl-ephedrine. Comment on the stereoselectivity of the two reactions.

44) (p. 62) Acetylation of (-)-2-octanol, C_6H_{13}-CHOH-CH_3 with acetic anhydride gives (-)-2-octyl acetate. Treatment of the same alcohol with p-toluenesulfonyl chloride and pyridine gives (-)-2-octyl p-toluenesulfonate which, when treated with sodium acetate in acetic acid, gives rise to (+)-2-octyl acetate. Explain.

45) (p. 62) (S)-(+)-hydratropic acid may be converted to (S)-(-)-α-phenethylamine as shown in Fig. 41. Treatment of the acid with PCl_3 followed by dimethyl cadmium gives (+)-α-phenethyl methyl ketone, $C_6H_5CH(CH_3)$-CO-CH_3. Treatment of this ketone with perbenzoic acid gives (-)-α-phenethyl acetate, $C_6H_5CH(CH_3)$-OCOCH$_3$. On the other hand, when (-)-α-phenethylamine (Fig. 41) is converted to the quaternary ammonium salt $C_6H_5CH(CH_3)N(CH_3)_3^+$ I^- which is then converted to the corresponding acetate by treatment with silver acetate and then pyrolyzed, the products are trimethylamine and (+)-α-phenethyl acetate. Explain.

46) How would you synthesize cis-stilbene, $C_6H_5CH=CHC_6H_5$ and trans-stilbene from 1,2-diphenylacetylene?

47) (pp. 62-65) What are the products of

a) Addition of diborane, B_2H_6, to 1-methylcyclohexene followed by oxidation with alkaline hydrogen peroxide?

b) Addition of bromine to cyclohexene?

c) Addition of hypobromous acid to trans-2-hexene followed by treatment with strong base to form the epoxide?

d) Direct epoxidation of trans-2-hexene with perbenzoic acid?

e) Ring opening of product from d) with hydrogen bromide?

f) Ring opening of product from d) with acidulated water?

g) Treatment of trans-3-hexene with hydrogen peroxide - osmium tetroxide?

h) Ring opening of cis-2-butene oxide with lithium aluminum deuteride?

48) (p. 65) Acetolysis of optically active threo-3-phenyl-2-butyl p-toluenesulfonate, CH_3-CHOTs-CH(C_6H_5)-CH_3 gives inactive threo-3-phenyl-2-butyl acetate free of erythro isomer. Explain.

49) (p. 66) Pyrolysis of <u>trans</u>-2-phenylcyclohexyl acetate gives mainly 1-phenylcyclohexene and acetic acid, but pyrolysis of the <u>cis</u> isomer gives largely 3-phenylcyclohexene and acetic acid.

50) (67-69) Illustrate the backbone of an isotactic, syndiotactic and heterotactic polymer. Are these polymer chains chiral? As obtained from synthesis of an α-olefin monomer, will they be optically active?

XI. ANSWERS TO PROBLEMS

1) a) Constitution. b) Configuration. c) Constitution. d) Constitution.
   e) Configuration. f) Configuration. g) Configuration. h) Constitution.
   i) Configuration.

2) Stereoisomers are isomers differing in spatial arrangement only.
   They must necessarily have the same constitution. It is generally
   held that stereoisomers differ in configuration, although some authors
   use the term also for molecular species which differ in conformation
   only.

3) Perhaps Pasteur's most important experimental discovery was the
   resolution of racemic modifications into (optically active) enantiomers.
   He found several methods to do this (see pp. 17-19), of which the
   method of mechanical separation is perhaps the most spectacular but
   not the most useful. Pasteur was the first scientist to realize clearly
   that optical activity was related to molecular dissymmetry (chirality).

4) a) A chiral center is a point in a molecule serving as a focus of
   dissymmetry (or chirality) by virtue of the arrangement of atoms or
   groups around it. In most cases (though not invariably) it is an atom
   to which different ligands are attached, for example a carbon atom of
   the type Cabcd (a, b, c and d being four different ligands). In lactic
   acid, $CH_3CHOHCOOH$, the #2 carbon is a chiral center by virtue of
   having four different ligands (CH_3, OH, H, COOH). b) Enantiomers
   are stereoisomers which bear a mirror-image relationship to each
   other, such as (+)- and (-)-tartaric acid (p. 5) or the dextro- and
   levorotatory lactic acids (p. 2). c) Racemic modifications are
   macroscopic assemblies of equal (or very nearly equal) numbers of
   the two enantiomers. Thus the 50:50 mixture of (+)- and (-)-α-bromo-
   propionic acids, $CH_3CHBrCOOH$, obtained in the bromination of
   propionic acid is a racemic modification.

5) Propionic acid, CH_3CH_2COOH, is achiral; α-hydroxypropionic or lactic acid, $CH_3CHOHCOOH$, is chiral. Lactic acid is also asymmetric but (-)-tartaric acid (p. 5) is an example of a chiral compound which is not asymmetric (it has a two-fold axis of symmetry). (-)-Tartaric acid may be called "dissymmetric" but the term "chiral" is now preferred.

6) a) Enantiomers. b) Diastereoisomers. c) Diastereoisomers. d) Enantiomers. e) Diastereoisomers. f) The terms are not normally applied to crystals, but if they were, one would have to consider the two types of crystals diastereoisomeric, since the unit cells are different. [This case is different from that of the sodium ammonium tartrate investigated by Pasteur which is a mechanical mixture of enantiomeric crystals of the (+)- and (-)-isomers.]

7) a) No. b) No. c) No. d) No. e) No. f) Yes. g) Yes. h) No. i) No. j) No. k) No. l) Yes. In general, enantiomers do not differ in "scalar" properties, i.e. properties which do not depend on absolute orientation in space.

8) Diastereoisomers differ from each other in the same way as constitutional isomers (e.g. 1-propanol and 2-propanol) do; they would therefore be expected to differ in all of the properties enumerated in question 7. The difference may be small in some cases (e.g. boiling point) and considerable in others (e.g. infrared spectrum), depending also, of course, on the particular pair of diastereoisomers considered.

9) a) Yes, those having a single chiral center, e.g. lactic acid (Fig. 1).
   b) Yes; in particular achiral cis-trans isomers, such as cis- and trans-4-methylcyclohexanol or cis- and trans-2-butene. (See also Fig. 4).
   c) Yes; in general, any compound with two or more chiral centers will be in this class, e.g. the tartaric acids (Fig. 2) or the 3-bromo-2-butanols (Fig. 5).

10) cis-trans Isomers are necessarily diastereoisomers, since they are stereoisomers which do not bear an object - mirror image relationship to each other. (Certain cis- or trans-forms, may, in addition, display enantiomerism as shown in Fig. 4, for example. However, this point is not pertinent to the question asked.)

11) a) See question 14a

b)

CH3
H-C-Br
H-C-Br
CH3

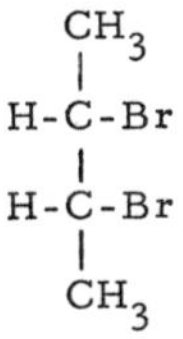

c)

CH3
H-C-Br
Br-C-H
CH3

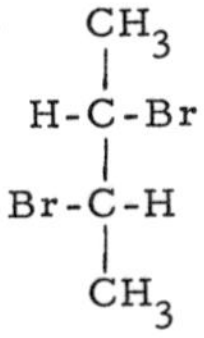

d)

CH3
H-C-OH
H-C-OH
C2H5

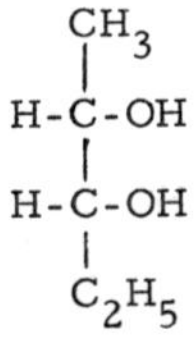

e)

CH3
H-C-OH
HO-C-H
C2H5

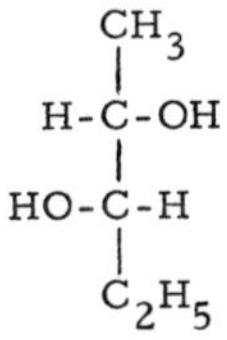

12)

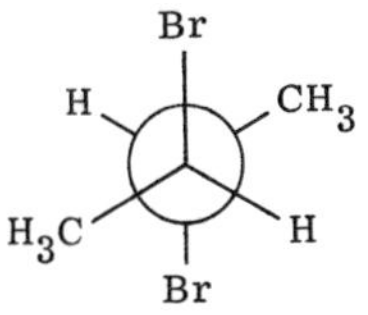

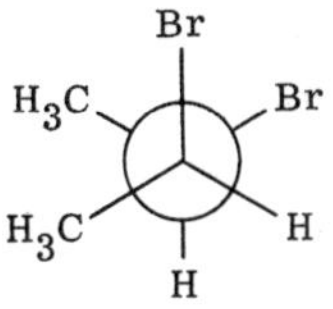

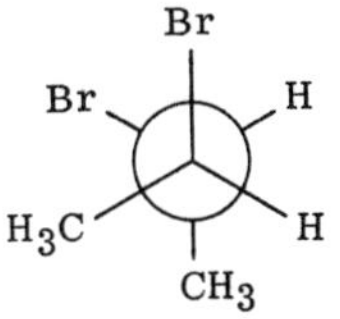

Br
H
CH3
H3C
H
Br

Br
H3C
Br
H3C
H
H

Br
Br
H
H3C
H
CH3

13)

c)

CH3
H
Br
H
Br
CH3
≡
H
Br
CH3
CH3
H
Br

d)

C2H5
H
OH
HO
H
CH3
≡
H
OH
C2H5
CH3
HO
H

e)

C2H5
H
OH
H
OH
CH3
≡
H
OH
C2H5
CH3
H
OH

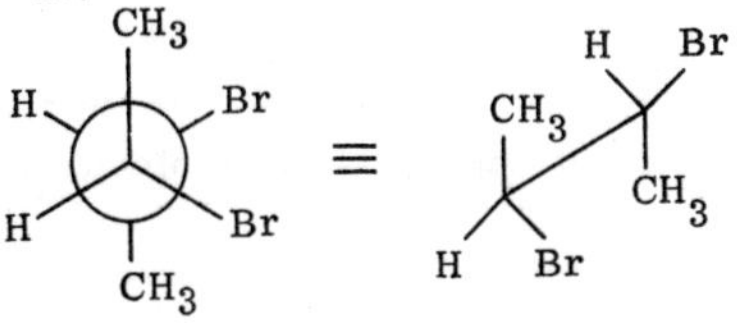

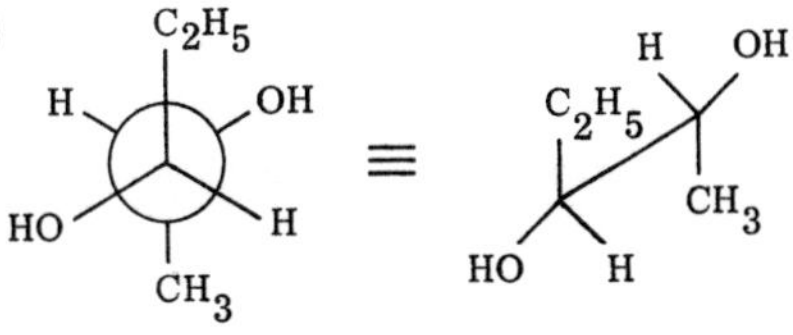

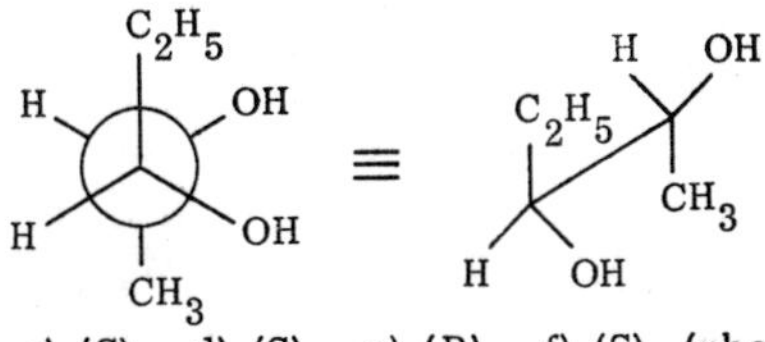

14) a) (S), b) (R), c) (S), d) (S), e) (R), f) (S) (phenyl precedes t-butyl). g) (2S, 3S) ($CHOHCH_2OH$ precedes CH_2OH). h) (2S, 3R) (this is a meso form).

15) a)   COOH   b)   COOH   c) CH_3 d)   OH   e) C_6H_4Cl

    H-C-OH H_2N-C-H   HO-C-H   D-C-H   H-C-OH

      CH_3 CH_3 C_6H_5 CH_3 C_6H_5

    f) See Fig. 2,   g) $CH=CH_2$ h)   OH

        p. 5. H_3C-C-H

                    CH_2CH_3

16) a) 4 (2 dl-pairs).   b) 8 (4 dl-pairs).   c) 8 (4 dl-pairs).

    d) 3 (1 dl-pair, 1 meso form).   e) 4 (1 dl-pair, 2 meso forms).

    f) 4 (2 dl-pairs).   g) 2 (cis and trans, both inactive).

    h) 3 (1 dl-pair, 1 meso form).   i) 2 (all-cis and cis-trans, both

    inactive).

17) No. A racemic modification is an assembly of a large number of
molecules of which very nearly 50% corresponds to one enantiomeric
form and very nearly 50% to the other. An individual molecule of a
chiral substance is either R or S, it cannot be racemic. If the substance is achiral, the term "racemic" is not applied to its molecules.

18) There is no difference in enthalpy, for, since the two enantiomers
have the same enthalpy, so has the racemic modification if it is an
ideal assembly of the two enantiomers. However, the racemic
modification, being composed of two different (enantiomeric) kinds
of molecules, is more randomous than either enantiomer by itself:
it has an excess entropy of mixing of R ln 2. Hence for the change
pure enantiomer → racemic modification the free energy decreases
by RT ln 2; i.e. the process is energetically favored.

19) No; approach from the two faces of the carbonyl function is equally
likely and a racemic modification will result (cf. Fig. 9). Again,
replacement of the two hydrogens by bromine is equally likely and a

racemic modification will again result. To obtain optically active mandelic acid, one must resolve the racemic modification, e. g. by crystallization of its ephedrine salt, or one must synthesize it in the presence of a chiral agent, e. g. by carrying out the HCN-addition in the presence of the enzyme emulsin.

20) The acylase enzyme is attuned to the hydrolysis of the acyl derivative of the naturally occurring (S-) amino acid. When it operates on racemic acylamino acid, only the derivative of the natural (S-) isomer is hydrolyzed and the unnatural (R-) isomer remains acylated. Incidentally, all the naturally occurring L-amino acids, save cysteine (cf. question 14e) are S.

21) a) (Malonic acid): equivalent.   b) (Isobutyric acid): enantiotopic. c) (Succinic acid): Hydrogen atoms on the same methylene group are enantiotopic. Hydrogen atoms on different methylene groups are pairwise equivalent (upper left with lower right; upper right with lower left).   d) (Pentane): The hydrogen atoms on the same carbon are enantiotopic; those at C-2 and C-4 are again pairwise equivalent. e) These hydrogen atoms are equivalent.   f) Enantiotopic. g) Equivalent. h) (Malic acid): diastereotopic.   i) (Citric acid): The hydrogen atoms on the same carbon are diastereotopic but those at C-2 and C-4 are pairwise enantiotopic (upper left enantiotopic with lower left; upper right enantiotopic with lower right). j) If n = 0, the answer is enantiotopic, if n〉0, diastereotopic. In all cases, the underlined hydrogen atoms are "heterotopic" (i. e. not equivalent or "homotopic"). k) Diastereotopic. l) Enantiotopic in either case.

22) a) Equivalent and enantiotopic protons are indistinguishable by nmr; diastereotopic protons ordinarily give distinct nmr signals, although there may be accidental coincidence. b) Heterotopic hydrogen atoms (whether enantiotopic or diastereotopic) are, in principle, distinct in reactions with chiral reagents, such as enzymes.

23) a) Enantiotopic. b) Equivalent. c) Diastereotopic. d) Enantiotopic
in so far as the two faces of each carbonyl group are concerned. However,
there is a pairwise equivalence between one of the two faces of each
carbonyl group. e) Equivalent. f) Enantiotopic.

24) a) Enantiomeric $C_6H_5CHOHCN$ species are formed. In ordinary
addition, these will be formed in identical amounts, but in enzyme-
catalyzed addition (e.g. in the presence of emulsin) one enantiomer
predominates. b) Same product. c) Two diastereoisomeric products
are formed (in unequal amounts). d) A single addition leads to two
enantiomeric products. In this product, the faces of the other carbonyl
function become diastereotopic, therefore a second addition of HCN
leads to two diastereoisomers. e) A single product. (Note this is
true only for cis-addition, for in trans-addition, the two atoms or groups
involved will not add to the same face of the double bond.) f) Two
enantiomeric products.

25) There are three eclipsed conformations (one Me/Me eclipsed, the
other two Me/H eclipsed) and three staggered ones (two gauche, one
anti). The case is entirely analogous to the 1,2-dibromoethane case
(Fig. 11, p. 25).

26) Refer to Fig. 11, p. 25 which shows the corresponding bromo compound.
Obviously not all the molecules can be in the anti conformation, or
else the dipole moment would be zero. Some molecules are in the
gauche conformation which has a finite dipole moment. This moment
may be calculated to be 3.2D. The fact that the observed moment is
only 1.12D indicates that the majority of molecules are in the anti form,
as expected.

27) The material crystallizes in a single conformation (anti; cf. Fig. 11,
p. 25) which, being centrosymmetric, shows relatively few ir absorption bands. On melting, equilibrium between the anti and gauche
conformations is established. Therefore the bands due to the anti
conformation become less intense and new bands, due to the gauche
conformation, appear.

28) For intramolecular hydrogen bonding to occur, the two hydroxyl
groups must be close together, i.e. _gauche_. For the _dl_-isomer,
one may write a conformation (A below) in which this is the case and,
at the same time, the methyl groups are in the more stable _anti_
position. For the _meso_ isomer, the conformations in which the
OH-groups are close to each other necessarily have the methyl
groups in the more crowded _gauche_ position (B); these conformations
will be less populated than conformation A of the _dl_-isomer.

29) a) Yes. b) Yes. c) No. The _trans_ acid cannot form an anhydride;
cf. Fig. 12, p. 29. d) Yes. e) No. f) No. Allenes show enantiomerism,
not _cis_-_trans_ isomerism. g) Yes. h) No.

30) For $CH_3CH=CHCH_3$ and $ClCH=CHCl$, the _cis_ isomer has the higher
dipole moment and therefore also the higher boiling point (which
correlates with dipole moment). For $CH_3CH=CHCl$, since the CH_3-C
and Cl-C dipoles are opposed, the _trans_ isomer has the higher dipole
moment and boiling point. Stability is not necessarily related to dipole
moment; as it happens, for 2-butene the _trans_ isomer is the more
stable and for the other two species, the _cis_ isomer. The reasons
for this order of stability are not completely understood; but it appears
that steric repulsion dominates in the case of _cis_-2-butene; whereas
for the more polarizable halogen compounds, the attractive London
forces dominate the situation in the _cis_ isomers.

31) "Small rings" have 3 or 4 members. They possess angle strain.
"Common rings" have 5, 6, or 7 members. The 5- and 7-membered
rings have eclipsing strain, the 6-membered rings are nearly strain-
free. "Medium rings" is the designation given to rings of 8-11 members.
They tend to have transannular or non-bonded strain due to steric interference of hydrogen atoms across the ring. They also have eclipsing

strain and they usually incur angle strain due to a widening of
the angles which occurs to minimize the non-bonded strain. In
contrast, the "large rings" (12 and up) have little if any strain.

32) The first two compounds will cyclize to form an oxirane (ethylene
oxide) more rapidly than the last two cyclize to form an oxetane
(trimethylene oxide) despite the greater strain in the three-membered
rings. This is because the probability of bringing the two ends of
a three-membered ring together is much greater than the corresponding probability for the four-membered ring. Because of the
Thorpe-Ingold effect (p. 35), the geminally dimethyl substituted
halohydrins ring-close more easily than the unsubstituted ones.
This is because the natural angle at $-C(Me)_2-$, 109.5°, is closer
to the ring angle (60 or 90°) than the angle at $-CH_2-$ (112.5°).

33) cis-Butene is less stable, (i.e. higher in enthalpy) than the trans
isomer, so its hydrogenation to the common product butane is more
exothermic. In the case of cyclononene, the stability is reversed:
the trans isomer is less stable than the cis and so its heat of hydrogenation to cyclononane is greater.

34) See Fig. 13, p. 31 and Fig. 14, p. 36.

35) a) No, the carboxy groups are too small to prevent rapid rotation
through the planar form at room temperature. b) Yes. c) No, the
molecule has two planes of symmetry even when the rings are at
right angles to each other. d) Yes, the large iodine atoms prevent
rapid rotation at room temperature. e) Yes. f) No, - see answer
to part c.

36) a) Yes, cis-trans isomers. b) Not normally. However, separation
of the two conformations has been effected at -150° (pp. 40-41). At
room temperature, the two conformations equilibrate at a rate of
nearly 10^5 sec^{-1}. c) Yes, cis-trans (or syn-anti) isomers. d) This
is one contributing resonance form of N-methylacetamide. The C-N
bond has much single bond character. Although the existence of cis-
trans isomers can be shown by nmr, isolation is not feasible in such

simple cases, but has been achieved in related instances and for similar thioamides (p. 40). e) Yes. f) No, see Fig. 21, p. 45. g) Yes, cis-trans isomerism; the cis is meso, the trans dl. h) No, because of rapid nitrogen inversion. However, cis-trans isomerism can be seen in the nmr spectrum (the isomers interconvert rapidly) and, in the case of the similar N-chloro compound, isomers have actually been separated (Fig. 21, p. 45).

37) Yes. (See Fig. 18, p. 41). The best resolving agent here is a platinum complex, Olefin · $PtCl_2 \cdot NH_2CH(CH_3)C_6H_5$. Model studies show that trans-cyclononene passes much more readily from one enantiomeric form into the other (by a ring inversion process) than does the eight-membered homolog; it cannot be resolved.

38) See Figs. 18-20, pp. 41, 43.

39) a) Quaternary ammonium salts are optically stable, but tertiary amines invert rapidly (p. 44). b) Abstraction of the α-hydrogen produces a carbanion which racemizes by an inversion process similar to that of a tertiary amine. When the carbanion

$$C_6H_5 \overset{O}{\overset{\|}{-C}} -\bar{C}(CH_3)C_2H_5$$ is reprotonated, a racemic mixture of the two

enantiomeric ketones results. c) Although the product is chiral, activity is lost at the stage of the intermediate radical, $C_6H_5 - \dot{C}D - CH_3$. d) Presumably the reaction passes through the carbonium ion

$$C_6H_5 - \overset{+}{\underset{\underset{CH_3}{|}}{C}} - C_6H_4OCH_3 - \underline{p}$$ which is planar.

40) i) See Fig. 22, p. 48. The case is similar to that of a spirane, Fig. 14, p. 36. ii) The case is completely analogous to that of a tetrahedral carbon with four different ligands, cf. Fig. 6, p. 12. iii) See Fig. 24, p. 49 (Pd in place of Pt). iv) See Fig. 25, p. 50 (Ni in place of Pt). v) See Fig. 23, p. 48 (for the general case). vi) See Fig. 27, p. 51, bottom part (Co for Pt and NO_2 for Cl). vii) See Fig. 30, p. 53 (Cr in place of Rh). viii) Four meso forms and one dl-pair:

$$
\begin{array}{c}
\text{NH}_3 \\
| \quad {.}\text{Cl} \\
\text{Br}-\text{Pt}-\text{Br} \\
\text{Cl} | \\
\text{NH}_3
\end{array}
\ (\underline{\text{meso}});
\qquad
\begin{array}{c}
\text{NH}_3 \\
| \quad {.}\text{Br} \\
\text{Br}-\text{Pt}-\text{Cl} \\
\text{Cl} | \\
\text{NH}_3
\end{array}
\ (\underline{\text{meso}});
\qquad
\begin{array}{c}
\text{NH}_3 \\
| \quad {.}\text{Cl} \\
\text{Br}-\text{Pt}-\text{Br} \\
\text{H}_3\text{N} | \\
\text{Cl}
\end{array}
\ (\underline{\text{meso}});
$$

$$
\begin{array}{c}
\text{NH}_3 \\
| \quad {.}\text{Br} \\
\text{Cl}-\text{Pt}-\text{Cl} \\
\text{H}_3\text{N} | \\
\text{Br}
\end{array}
\ (\underline{\text{meso}});
\qquad
\begin{array}{c}
\text{NH}_3 \\
| \quad {.}\text{Br} \\
\text{Br}-\text{Pt}-\text{Cl} \\
\text{H}_3\text{N} | \\
\text{Cl}
\end{array}
\ \text{and}
\begin{array}{c}
\text{NH}_3 \\
| \quad {.}\text{Br} \\
\text{Cl}-\text{Pt}-\text{Br} \\
\text{H}_3\text{N} | \\
\text{Cl}
\end{array}
\ (\underline{\text{dl}}\text{-pair}).
$$

ix) Two <u>meso</u> forms and three <u>dl</u>-pairs:

meso   meso   <u>dl</u>-pair

    <u>dl</u>-pair   <u>dl</u>-pair

41) Regarding the two possible configurations of the starting material, see Fig. 27 (p. 51), top part (NH$_3$ in place of Cl and NO$_2$ in place of NH$_3$). Clearly the <u>trans</u> isomer (NH$_3$ groups <u>trans</u>) can give only a single, optically inactive product by replacement of two (adjacent) NO$_2$ groups by H$_2$NCH$_2$CH$_2$NH$_2$. The <u>cis</u> isomer, however, can give rise to two different products, one of which is resolvable:

$$
\begin{array}{c}
NO_2 \\
| \quad .N \\
H_3N-Co-N \\
H_3N \quad | \\
NO_2
\end{array}
\qquad
\begin{array}{c}
N \\
| \quad N \\
H_3N-Co-NO_2 \\
H_3N \quad | \\
NO_2
\end{array}
\quad \text{and} \quad
\begin{array}{c}
NO_2 \\
| \quad .N \\
H_3N-Co-NO_2 \\
H_3N \quad | \\
N
\end{array}
$$

<u>meso</u>   <u>dl</u>-pair

42) See Fig. 36, p. 58. That the rate of racemization is faster than
the rate of radioactive oxalate ion exchange with the complex is
proof that complete detachment of an oxalate group is not required
for racemization. The fact that all twelve of the oxygen atoms in the
complex exchange during racemization indicates that the Cr-O
bonds must be ruptured making all oxygen atoms equivalent - otherwise only six would exchange. That the rate of oxygen exchange
is similar to the rate of racemization implies that not all chelate
ring opening-closing acts lead to racemization.

43) (See Eliel, Stereochemistry of Carbon Compounds, p. 435). The
catalytic reduction of the hydrochloride over platinum is highly
stereoselective. The chemical reaction is only very weakly selective
and the selectivity is in the direction of the less desired isomer.

44) (See Eliel, Stereochemistry of Carbon Compounds, p. 117). Acetylation
of (R)-(-)-octanol gives the (R)-(-)-acetate of unchanged configuration.
Conversion of the (-)-alcohol to (R)-(-)-2-octyl p-toluenesulfonate
also leaves configuration unchanged, but reaction of the toluenesulfonate with acetate involves inversion (cf. Fig. 40, p. 62, except
that toluenesulfonate is the leaving group instead of bromide) and
thus gives the (S)-acetate which is dextrorotatory.

45) The rearrangement of the ketone to the acetate (Baeyer-Villiger
reaction) involves retention, similar to the Hofmann rearrangement
(Fig. 41); therefore the configuration of (-)-α-phenethyl acetate
obtained in this rearrangement corresponds to that of the (-)-amine

(Fig. 41, -OAc in place of $-NH_2$), i.e. it is S. However, pyrolysis of the quaternary ammonium salt involves inversion and thus leads to the (R)-(+)-acetate:

$$
\begin{array}{ccc}
\underset{|}{\overset{\displaystyle CO-CH_3}{}} & & \underset{|}{\overset{\displaystyle OCOCH_3}{}} \quad \underset{|}{\overset{\displaystyle CH_3}{}} \\
H_3C-\underset{|}{C}-H & \xrightarrow{C_6H_5CO_3H} & H_3C-\underset{|}{C}-H \quad\equiv\quad H-\underset{|}{C}-OCOCH_3 \\
C_6H_5 & & C_6H_5 \qquad\quad C_6H_5
\end{array}
$$

$$\text{(S)-(+)}\qquad\qquad\qquad\qquad\qquad\text{(S)-(-)}$$

$$
\begin{array}{ccc}
\underset{|}{\overset{\displaystyle CH_3}{}} & & \underset{|}{\overset{\displaystyle CH_3}{}} \\
H-\underset{|}{C}-N(CH_3)_3^+\ OAc^- & \xrightarrow{\;\delta\;} & AcO-\underset{|}{C}-H\ +\ N(CH_3)_3 \\
C_6H_5 & & C_6H_5
\end{array}
$$

$$\text{(S)}\qquad\qquad\qquad\qquad\text{(R)-(+)}$$

46) Catalytic reduction over palladium can be stopped at the olefin stage and gives largely cis-stilbene. Chemical reduction (sodium, liquid ammonia) yields trans-stilbene.

47) a) This is a cis-addition and therefore gives trans-2-methylcyclo-hexanol (hydrogen and OH cis). b) trans-1, 2-Dibromocyclohexane since the cyclic olefin is necessarily cis and addition is trans or anti. c) anti Addition yields a mixture of the erythro 2, 3-bromydrins (2-Br, 3-OH and 3-Br,2-OH); ring closure with inversion then yields the trans epoxide. d) This, also, yields the trans epoxide by a direct cis-addition. e) The ring opening proceeds with inversion of configuration at C-2 or C-3 and yields the same bromohydrin mixture as c). f) This, also, proceeds with inversion and therefore gives erythro-2, 3-hexanediol. g) threo-2, 3-Hexanediol (cis addition).
h)

$$
\begin{array}{l}
\qquad\ \overset{\displaystyle CH_3}{\underset{|}{}} \\
\ H-\underset{|}{C}-D \\
HO-\underset{|}{C}-H \\
\qquad CH_3
\end{array}
$$
and its enantiomer.

(Note that all products obtained in question 47 are racemic modifications.)

48) The reaction appears to proceed by an intermediate phenonium
ion; the case is similar to that depicted in Fig. 44, p. 65:

$$
\begin{array}{ccc}
\underset{\displaystyle CH_3}{\overset{\displaystyle CH_3}{\underset{H_5C_6-\overset{|}{\underset{|}{C}}-H}{H-\overset{|}{\underset{|}{C}}-OTs}}} & \longrightarrow & \text{(symmetric)} & \longrightarrow & \underset{\displaystyle CH_3}{\overset{\displaystyle CH_3}{\underset{H_5C_6-\overset{|}{\underset{|}{C}}-H}{H-\overset{|}{\underset{|}{C}}-OAc}}} + \underset{\displaystyle CH_3}{\overset{\displaystyle CH_3}{\underset{H-\overset{|}{\underset{|}{C}}-OAc}{H_5C_6-\overset{|}{\underset{|}{C}}-H}}}
\end{array}
$$

symmetric   (d̲l̲-pair)

49) When it is stereochemically allowed, the reaction yields the more
stable conjugated olefin, 1-phenylcyclohexene. However, the
reaction requires c̲i̲s̲-elimination, similar to amine oxide pyrolysis
(Fig. 45, p. 66) and in c̲i̲s̲-2-phenylcyclohexyl acetate, the only
available c̲i̲s̲ hydrogen is at C-3, so the elimination goes in that
direction.

50) See Fig. 46. The heterotactic chain is, in principle, chiral whereas
isotactic and syndiotactic chains would appear to have a symmetry
plane through an R-C-H link (and, in the case of the isotactic chain,
also through a CH_2 link) near the middle of the chain if one disregards
small irregularities near the chain ends. In any case, synthesis
from an achiral olefin precursor with a symmetric catalyst necessarily gives an inactive polymer.

X. Index